Designing Robot Behavior
in Human-Robot Interactions

Changliu Liu
Robotics Institute, Carnegie Mellon University
Pittsburgh, Pennsylvania, USA

Te Tang
Advanced Research Laboratory
FANUC America Corporation, Union City, California, USA

Hsien-Chung Lin
Advanced Research Laboratory
FANUC America Corporation, Union City, California, USA

Masayoshi Tomizuka
Department of Mechanical Engineering
University of California, Berkeley, California, USA

CRC Press
Taylor & Francis Group
Boca Raton London New York

CRC Press is an imprint of the
Taylor & Francis Group, an **informa** business
A SCIENCE PUBLISHERS BOOK

CRC Press
Taylor & Francis Group
6000 Broken Sound Parkway NW, Suite 300
Boca Raton, FL 33487-2742

First issued in paperback 2020

ISBN-13: 978-0-367-17969-4 (hbk)
ISBN-13: 978-0-367-77657-2 (pbk)

Library of Congress Cataloging-in-Publication Data

Names: Liu, Changliu, 1990- author. | Tang, Te, author. | Lin, Hsien-Chung, author. | Tomizuka, M., author.
Title: Designing robot behavior in human-robot interactions / Changliu Liu, Robotic Institute, Carnegie Mellon University, Pittsburgh, Pennsylvania, USA, Te Tang, Department of Mechanical Engineering, University of California, Berkeley, USA, Hsien-Chung Lin, Research Engineer at FANUC America Corp, San Francisco, California, USA, Masayoshi Tomizuka, Department of Mechanical Engineering, University of California, Berkeley, USA.
Description: Boca Raton : CRC Press, Taylor & Francis Group, [2019] | "A science publishers book." | Includes bibliographical references and index. | Summary: "In this book, we have set up a unified analytical framework for various human-robot systems, which involve peer-peer interactions (either space-sharing or time-sharing) or hierarchical interactions. A methodology in designing the robot behavior through control, planning, decision and learning is proposed. In particular, the following topics are discussed in-depth: safety in the process of during human-robot interactions, efficiency in real-time robot motion planning, imitation of human behaviors from demonstration, dexterity of robots to adapt to different environments and tasks, cooperation among robots and humans with conflict resolution. These methods are applied in various scenarios, such as human-robot collaborative assembly, robot skill learning from human demonstration, interaction between autonomous and human-driven vehicles, etc"-- Provided by publisher.
Identifiers: LCCN 2019026334 | ISBN 9780367179694 (hardcover : acid-free paper)
Subjects: LCSH: Human-robot interaction. | Robotics.
Classification: LCC TJ211.49 .L58 2019 | DDC 629.8/924019--dc23
LC record available at https://lccn.loc.gov/2019026334

Visit the Taylor & Francis Web site at
http://www.taylorandfrancis.com

and the CRC Press Web site at
http://www.crcpress.com

Preface

Human-robot interactions (HRI) have been recognized to be a key element of future robots in many application domains such as manufacturing and transportation, which have huge social and economic impacts. Future robots are envisioned to be acting like humans, which are independent entities that make decisions for themselves; intelligent actuators that interact with the physical world; and involved observers that have rich senses and critical judgements. Most importantly, they are entitled social attributions to build relationships with humans. We call them co-robots.

Technically, it is challenging to design the behavior of co-robots. Unlike traditional robots which work in structured and deterministic environments, co-robots need to operate in highly unstructured and stochastic environments. The book is dedicated to study the methodologies to ensure that co-robots operate efficiently and safely in dynamic uncertain environments.

This book sets up a unified analytical framework for various human-robot systems, which involve peer-peer interactions or hierarchical interactions. Various methods to design robot behavior through control, planning, decision, and learning are proposed. In particular, the following topics are discussed: **safety** during human robot interactions, **efficiency** in real-time robot motion planning, **imitation** of human behaviors by robot, **dexterity** of robots to adapt to different environments and tasks, **cooperation** among robots and humans and conflict resolution. The proposed methods have been applied on various scenarios, such as human-robot collaborative assembly, robot skill learning from human demonstration, interaction between autonomous and human-driven vehicles, etc.

Acknowledgement

This book is based on the doctoral researches of Prof. Liu, Dr. Tang, and Dr. Lin in Prof. Tomizuka's group at University of California, Berkeley. A very sincere gratitude goes out to all the collaborators on various topics:

1. Dr. Wei Zhan on speed profile planning in Chapter 4;
2. Dr. Yu Zhao and Dr. Wenjie Chen on force control by imitation learning in Chapter 5;
3. Mr. Changhao Wang on the framework of analogy learning in Chapter 6;
4. Prof. Wenlong Zhang on Blame-All strategy in Chapter 7;
5. Prof. Chung-Wei Lin and Dr. Shinichi Shiraishi on distribution conflict resolution in Chapter 7;
6. Mr. Jianyu Chen and Dr. Trong-Duy Nguyen on ROAD in Chapter 8;
7. Dr. Yongxiang Fan on grasp planning by analogy learning in Chapter 9;
8. Ms. Yujiao Cheng on SERoCS in Chapter 10.

The authors are very grateful to all the sponsors of the works presented in this book, including the National Science Foundation, and industrial sponsors: FANUC Corporation, Denso International America, and Toyota InfoTechnology Center. Their support and insightful discussions have enriched the results of this book.

Finally, the authors would like to thank many of their colleagues, students, and friends for their precious suggestions. In particular, we would like to thank Prof. Xu Chen, Mr. Weiye Zhao, Mr. Jaskaran Grover, and Mr. Angelos Mavrogiannis.

Contents

Preface   iii

Acknowledgement   v

Nomenclature   xiii

PART I INTRODUCTION

1. Introduction   3

   1.1 Human-Robot Interactions: An Overview   3
   1.2 Modes of Interactions   4
   1.3 Robot Behavior System   7
      1.3.1 Overview   7
      1.3.2 Knowledge   11
      1.3.3 Logic   14
      1.3.4 Learning Process   16
   1.4 Design Objectives   17
      1.4.1 Safety   17
      1.4.2 Efficiency   18
      1.4.3 Imitation   19
      1.4.4 Dexterity   20
      1.4.5 Cooperation   20
   1.5 System Evaluation   21
      1.5.1 Theoretical Evaluation 21
      1.5.2 Experimental Evaluation 22
   1.6 Outline of the Book   24

2. Framework   27

   2.1 Multi-Agent Framework   27
      2.1.1 The General Multi-Agent Model   27
      2.1.2 Models for Different Modes of Interactions 28
      2.1.3 Features of Human-Robot Systems   30
   2.2 Agent Behavior Design and Architecture   31
      2.2.1 Model-Based Design   31
      2.2.2 Model-Free Design   37
   2.3 Conclusion   41

PART II THEORY

3. **Safety during Human-Robot Interactions**   **45**

    3.1   Overview   45
    3.2   Safety-Oriented Behavior Design   46
    3.3   Safe Set Algorithm   49
        3.3.1   The Algorithm   49
        3.3.2   Example: Planar Robot Arm   50
        3.3.3   Example: Vehicle   53
    3.4   Safe Exploration Algorithm   54
        3.4.1   The Algorithm   54
        3.4.2   Example: Local Planning of a Vehicle   56
    3.5   An Integrated Method for Safe Motion Control   59
    3.6   Conclusion   59

4. **Efficiency in Real-Time Motion Planning**   **61**

    4.1   Overview   61
    4.2   Problem Formulation   62
    4.3   Optimization-Based Trajectory Planning   64
        4.3.1   Problem Formulation   65
        4.3.2   Quadratic Approximation   66
        4.3.3   Examples: Trajectory Planning for Various Systems   68
    4.4   Optimization-Based Speed Profile Planning   71
        4.4.1   Problem Formulation   72
        4.4.2   Quadratic Approximation   75
        4.4.3   Example: Vehicle Speed Profile Planning   77
    4.5   Optimization-Based Layered Planning   78
    4.6   Conclusion   81

5. **Imitation: Mimicking Human Behavior**   **83**

    5.1   Overview   83
    5.2   Imitation for Prediction   83
        5.2.1   Classification of Behaviors   84
        5.2.2   Adaptation of Behavior Models   86
    5.3   Imitation for Action   89
        5.3.1   Imitation Learning by Gaussian Mixture Model   90
        5.3.2   Interpreting GMR from Mechanics, Point of View   92
        5.3.3   Stability Condition of the Closed-Loop System   93
    5.4   Conclusion   96

6. Dexterity: Analogy Learning to Expand Robot Skill Sets   **97**

6.1 Overview   97
6.2 Concept of Analogy Learning   97
6.3 Advantages of Analogy Learning   98
    6.3.1 Learn from SMALL Data   99
    6.3.2 Robustness to Input Noise   100
    6.3.3 Guaranteed Collision-Free Property   101
6.4 Structure Preserved Registration for Analogy Learning   103
    6.4.1 Node Registration with Gaussian Mixture Model   103
    6.4.2 Regularization on Local Structure   105
    6.4.3 Regularization on Global Topology   108
    6.4.4 Closed-form Solution for Transformation Function   108
    6.4.5 Implementation Acceleration   113
6.5 Experimental Study   113
6.6 Conclusion   117

7. Cooperation: Conflict Resolution during Interactions   **119**

7.1 Overview   119
7.2 Dynamics of Multi-Agent Systems   121
    7.2.1 Simultaneous Dynamic Game   121
    7.2.2 Agent Strategies   122
    7.2.3 Multi-Agent Learning   123
    7.2.4 Trapped Equilibrium   124
7.3 Cooperation under Information Asymmetry   125
    7.3.1 Quadratic Game and Nash Equilibrium   125
    7.3.2 Blame-Me Strategy   128
    7.3.3 Blame-All Strategy   129
    7.3.4 Example: Robot-Robot Cooperation   131
7.4 Conflict Resolution through Communication   133
    7.4.1 Conflict Resolution and Nash Equilibria   134
    7.4.2 Communication Protocol and Strategy   136
    7.4.3 Example: Unmanaged Intersections   144
7.5 Conclusion   146

PART III APPLICATIONS

8. Human-Robot Co-existence: Space-Sharing Interactions   **149**

8.1 Overview   149
8.2 Robot Safe Interaction System for Industrial Robots   150
    8.2.1 The Optimization Problem   151
    8.2.2 The Robot Safe Interaction System   152

8.2.3   Case Studies   156
8.2.4   Discussion and Conclusion   159
8.3   Robustly-Safe Automated Driving System   159
8.3.1   Multi-Agent Traffic Model   160
8.3.2   Prediction and Planning in the ROAD System   162
8.3.3   Case Studies   168
8.3.4   Discussion and Conclusion   172
8.4   Conclusion   174

9. Robot Learning from Human: Hierarchical Interactions   **175**

9.1   Overview   175
9.2   Remote Lead Through Teaching for Implementing   175
    Imitation Learning
9.2.1   Human Demonstration Phase   175
9.2.2   Skill Learning Process   176
9.2.3   Robot Reproduction Phase   177
9.2.4   Experimental Study   177
9.3   Robotic Grasping by Analogy Learning   180
9.3.1   Background of Robotic Grasping   181
9.3.2   Grasp Planning with Human Demonstration   182
9.3.3   Grasping Pose Transferring by Analogy Learning   183
9.3.4   Experimental Study   187
9.4   Robotic Motion Re-Planning by Analogy Learning   190
9.4.1   Path Re-Planning by Analogy Learning   192
9.4.2   Trajectory Re-Planning by Analogy Learning   195
9.5   Conclusion   198

10. Human-Robot Collaboration: Time-Sharing Interactions   **201**

10.1 Human-Robot Collaboration in Manufacturing   201
10.2 Safe and Efficient Robot Collaborative System   202
10.2.1 Mathematical Problem   202
10.2.2 Architecture of SERoCS   203
10.2.3 Example   204
10.3 Experimental Study   204
10.3.1 Experiment Setup   204
10.3.2 Validation of the Environment Monitoring   206
10.3.3 Robots in the Idle Mode   207
10.3.4 Human-Robot Collaborative Assembly   209
10.4 Conclusion   213

PART IV Conclusion

11. Vision for Future Robotics and Human-Robot Interactions   **217**

    11.1 Roadmap for the Future   217
        11.1.1 The Demand Side   217
        11.1.2 Future Robotics   218
        11.1.3 Human-Robot Relationships   219
        11.1.4 Ethical Issues   219
    11.2 Conclusion of the Book   220

References   **221**

Index   **235**

Color Section   **237**

Nomenclature

x	state
u	control
y	measurement
π	historical data (combination of all measurements)
t	time
k	time step
t_s	sampling time
g	logic function (from measurement to control)
f	dynamic function (from control to state)
h	measurement function (from state to measurement)
$\mathcal{J}$	cost
$\mathcal{M}$	model
$\mathcal{B}$	behavior system (including knowledge, logic, and learning)
$\mathcal{X}$	state space
$\mathcal{C}$	configuration space
Ω	constraint on control space
Γ	constraint on state space
$\mathcal{I}$	constraint for interaction
ϕ	safety index
$\mathbb{E}$	expectation
$\mathbb{P}$	probability
$\mathcal{N}$	normal distribution
κ	curvature
$\hat{}$	estimate
$\tilde{}$	estimation error
X	mean squared estimation error
$\mathbf{I}_n$	$n \times n$ identity matrix
w	contact wrench
K	controller gain
$\varnothing$	empty set
σ	covariance
Σ	covariance matrix
$\mathcal{T}$	registration function
$\mathcal{O}$	complexity
$.^\top$	matrix transform
$\bar{}$	registration point

Part I

Introduction

1 Introduction

1.1 HUMAN-ROBOT INTERACTIONS: AN OVERVIEW

Human-robot interactions (HRI) have been recognized to be a key element of future robots in many application domains such as manufacturing, transportation, service, and entertainment. In factories, robots are expected to collaborate with human workers [109]. Manufacturers are interested in combining human's flexibility and robot's productivity in flexible production lines [102]. In transportation systems, autonomous vehicles are expected to interact with various road participants, including human-driven vehicles and pedestrians. In hospitals, medical robots are expected to help doctors to perform surgeries on patients. Exoskeleton robots are expected to assist stroke patients to walk [100]. Nursing robots [169] are expected to take care of patients. Robot guide dogs [203] are expected to help blind people.

These applications entail huge social and economic impacts [52, 34]. Future robots are envisioned to be acting like humans, which are independent entities that make decisions for themselves; intelligent actuators that interact with the physical world; and involved observers that have rich senses and critical judgements. Most importantly, they are entitled social attributions to build relationships with humans [63]. We call these robots co-robots.

Technically, it is a challenge to design the behavior of co-robots. Unlike conventional robots that work in structured and deterministic environments, co-robots need to operate in highly unstructured and stochastic environments. The skill sets that have been developed for conventional robots can no longer meet the needs of co-robots. This book explores methodologies to equip co-robots with desired skill sets, in order to enable wide adoption of co-robots in various interactive environments, as mentioned earlier. The fundamental research question to address in this book is *how to ensure that co-robots operate efficiently and safely in dynamic uncertain environments.* Regarding broadness and complexity of HRI [52], the following four aspects of the fundamental question are considered:

Diverse modes of interactions

The potential applications of co-robots lie in various domains, with different modes of interactions [167, 111, 55]. For example, humans and robots can be collaborators in factories, while automated vehicles and vehicles driven by human drivers are competitors for limited road resources on public roads. Different modes of interactions imply different system dynamics as well as diverse performance requirements for the robot. Moreover, the relationship that a robot builds with a human may be dynamic, i.e., the mode of interaction may change over time. A unified model for various HRI applications is needed in order to provide a comprehensive understanding of HRI, guide the design of the robot behavior, and serve as an analytical framework for performance evaluation of the human-robot systems.

Design of the robot behaviors

Behavior is the way in which one acts or conducts oneself, especially towards others. Action can be taken in multiple ways. For a humanoid robot, it can be physical movement, gesture, facial expression, language, etc. For an automated vehicle, it can be physical movements, turning on lights, honking, etc. We study the methodology of behavior design, i.e., how to realize the design goal (to ensure that co-robots operate efficiently and safely in dynamic uncertain environments) within the design scope (the inputs and outputs of the robotic system). This book mainly considers physical movements of co-robots.

Software embodiment of the behavior and its computation efficiency

The designed behavior is embodied as robot software. The complexity of the software will increase dramatically when the environment or the task becomes more complex. To ensure timely responses to environmental changes and guarantee safety during operation, real-time computation and actuation is crucial. Efficient algorithms are highly demanded.

Analysis, synthesis and evaluation of complex human-robot systems

The effectiveness of the designed robot behavior needs to be evaluated in human-robot systems. The evaluation can be performed theoretically as well as experimentally. The challenge in theoretical analysis is the complication of interweaving software modules. The difficulty in conducting experiments is that the tolerance of failure is extremely low when human subjects are in the loop, i.e., human safety is critical. Moreover, it is difficult to repeat some human-in-the-loop experiments due to the stochasticity of human behaviors. It is important to develop effective evaluation platforms for human-robot systems.

The focus of this book is to (1) set up a unified analytical framework for various human-robot systems; (2) establish methodologies to design robot behaviors under the framework that address the fundamental question. The considerations in the above four aspects will be elaborated in the following sections.

1.2   MODES OF INTERACTIONS

Interaction between a human and a robot may have various modes. We divide it into two kinds of relationships: *parallel relationships* and *hierarchical relationships*.

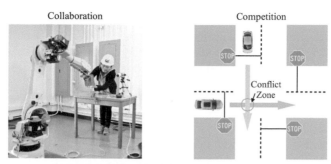

(a) Parallel relationships in human-robot interactions.

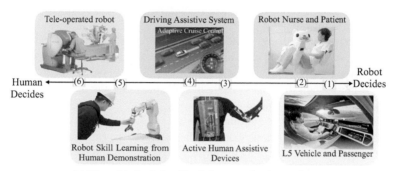

(b) Hierarchical relationships in human-robot interactions.

Figure 1.1: Various modes of human-robot interactions.

Color version at the end of the book

Parallel HRI

In a parallel relationship, a human and a robot are two independent entities who make their own decisions. In literature, parallel interaction is also called peer-peer interaction [55]. Typical examples of parallel relationships are (1) the interaction between an automated vehicle and a human-driven vehicle and (2) the interaction between an industrial co-robot and a human worker in production lines. In either case, the robot (the automated vehicle or the industrial co-robot) and the human (the human-driven vehicle or the human worker) are peers as opposed to hierarchical relationships. In the parallel relationships shown in Fig. 1.1a, the actions of the human and the robot either need to be synchronized (e.g., when the human and the robot are moving one workpiece cooperatively) or need to be asynchronized (e.g., two vehicles cannot occupy the conflict zone at the same time when they are crossing an intersection). We call the synchronized actions *collaboration* and the asynchronized actions *competition* (since there will always be a vehicle that passes the conflict zone first). Competition is the most common interaction mode. For instance, if a human and a robot are competing for space, the competition can be understood as collision avoidance.

Hierarchical HRI

In hierarchical relationships, either the human or the robot transfers the responsibility of making decisions to the other, completely or partially. Typical examples in hierarchical relationships are listed below, which is also shown in Fig. 1.1b.

1. The interaction between an automated vehicle and a passenger inside a vehicle, where the human passenger allows the vehicle to make driving decisions.[1]
2. The interaction between a robot nurse and a patient, where the robot decides the moving trajectory for the patient.[2]
3. The interaction between a human and a human-assistive device such as an exoskeleton. The human can be guided by the robot, but can also "fight" against the robot.
4. The interaction between a human driver and a driver-assistance system.[3] The driver-assistance system can serve as a guardian angel [89] that allows the driver to decide in safe situations and takes over during an emergency, or as an assistant that takes charge in safe situations and asks the human to take charge of emergencies.[4]
5. The interaction between a human and a robot when the human teaches the robot a skill by kinesthetic teaching. In this case, the robot is following the moving trajectory taught by the human.
6. The interaction between a human tele-operator and a tele-operated robot, where the robot purely follows the command from the human.[5]

As discussed in the above examples, the allocation of responsibilities varies for different hierarchical interactions. The robot purely follows the human's command in one extreme, while the human purely follows the robot's motion in another extreme. When human dominates the decision process, it is called supervisory interaction [153] in literature. Allocation of responsibility can change throughout time as discussed in the guardian angel case for driver-assistance systems.

The modes of interactions among multiple humans and robots can be built from those basic interaction modes between one human and one robot. A multi-agent framework will be proposed in Chapter 2 to provide a unified framework to analyze various kinds of interactions, where every intelligent entity (human or robot) is viewed as one agent. For simplicity, we do not distinguish between a robot and a co-robot literally in the following discussion.

---

[1]Online figure: http://www.autonews.com/article/20160111/OEM06/301119965/autonomous-vehicle-architects-begin-to-contemplate-the-human-inside.

[2]Online figure: http://newatlas.com/riba-robot-nurse/12693/.

[3]Online figure: http://editorial.autoweb.com/autowebs-guide-to-adaptive-cruise-control/.

[4]The level of autonomy of an intelligent vehicle is divided into six categories in the SAE International's new standard J3016 [6], which ranges between the two extremes: either the human driver makes all the decisions or the automated vehicle makes all the decisions.

[5]Online figure: http://www.telepresenceoptions.com/2014/08/r/.

1.3 ROBOT BEHAVIOR SYSTEM

This book focuses on behavior design of a robot from the perspective of physical movement, e.g., how to generate safe and efficient motions during interactions.

1.3.1 OVERVIEW

Three components

To generate desired robot behavior, we need to

1. encode correct knowledge in the robot;
2. design a correct logic to let the robot turn the knowledge into desired actions;
3. design a learning process to update the knowledge and the logic, in order to make the robot adaptable to unforeseen environments.

Knowledge, *logic*, and *learning* are the major components of a behavior system, as shown in Fig. 1.2a. In the diagram, the robot obtains measurement π from the human-involved environment and generates its action u according to the logic.

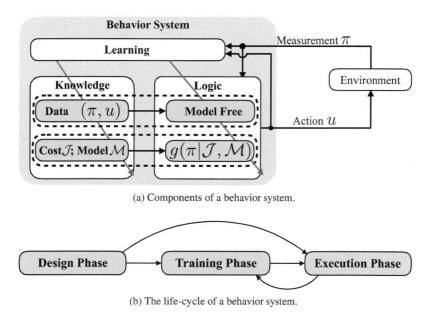

(a) Components of a behavior system.

(b) The life-cycle of a behavior system.

Figure 1.2: Behavior systems.

Knowledge and logic can be considered in both the model-based framework or the model-free framework. Model-free knowledge consists of a set of data points

(π, u), which implicitly encodes the relationship between measurement and action. The corresponding logic is a model-free logic that acts according to the implicit mapping from π to u observed in the data points. Model-based knowledge consists of *cost* that describes task requirements and *model* that describes the dynamics of the environment. The corresponding logic $g(\pi | \mathcal{J}, \mathcal{M})^6$ is a mapping from measurement to action that depends on its knowledge: the cost $\mathcal{J}$ and the model $\mathcal{M}$.

The learning process updates the knowledge and the logic based on the data π, which is necessary since (1) the designed knowledge may not cover all possible scenarios and (2) the environment may be time varying. The mathematical formulation will be further explained in Chapter 2.

Robot lifetime

The lifetime of a robot is divided into three phases: the design phase, the training phase and the execution phase, as shown in Fig. 1.2b. We call the first two phases offline process and the third online process. In the design phase, the human designers specify the three components in Fig. 1.2a for the robot. In the training phase, the robot can learn knowledge from experience or from human demonstration. The difference between the knowledge learned from human demonstration and the knowledge designed by human is that the former does not require the human to have a mathematical or quantitative representation of the knowledge. In many cases, such mathematical representations are hard to obtain and unintuitive for the human. For example, it is easier for a human to gesture a motion trajectory than to come up with a mathematical function of the trajectory. In the execution phase, the robot performs its task and interacts with the environment. While performing the tasks, the robot can update the knowledge or the logic through online learning. However, due to limitations in computation power, online learning is restricted to small-scale parametric adaptation. Structural changes such as learning a new skill from scratch are performed by offline learning in the training phase. The training phase and the execution phase can be performed iteratively in a lifelong learning system [212]. It is also possible that the robot goes directly from the design phase to the online execution phase without going through the training phase.

Nature vs. nurture

Knowledge is the key content of a behavioral system. How much it should be designed (nature) and how much it should be learned (nurture) still remains arguable [51].

Although knowledge can be learned, the two other components, logic and learning, are essentially algorithms to be designed. There are three ways to obtain logic g as illustrated in Fig. 1.3. The contours in the figures represent the internal cost $\mathcal{J}(u, \pi)$. The darker the color, the higher the cost. Logic g is a mapping from π to u, and can be obtained using the following three methods. The first two methods

---

^{6}The function g is also called a control law in classic control theory or a control policy in decision theory.

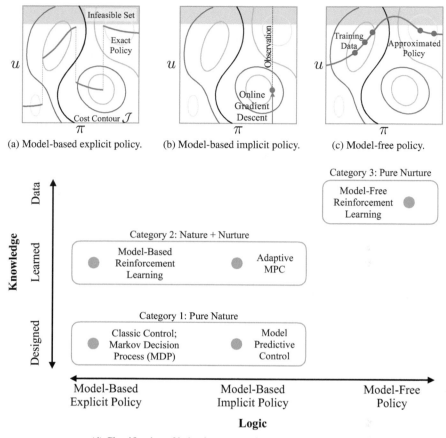

(a) Model-based explicit policy.   (b) Model-based implicit policy.   (c) Model-free policy.

(d) Classification of behavior systems from nature to nurture.

Figure 1.3: Classification of behavior systems.

are model-based, which assume explicit $\mathcal{J}$ and $\mathcal{M}$. The last method is model-free, which only requires (π, u) data points.

1. Model-based explicit policy.
   We can explicitly solve the optimization $\min_u \mathcal{J}(u, \pi)$ given $\mathcal{M}$ in the design phase (offline). In this way, we obtain a model-based *explicit policy* $g : \pi \mapsto u$ which solves the optimization, i.e., $g(\pi) := \min_u \mathcal{J}(u, \pi)$, as shown in the thick curve in Fig. 1.3a. The function g may be discontinuous, when the cost $\mathcal{J}$ is non-convex.

2. Model-based implicit policy.
   The optimization can also be solved online in the execution phase. An algorithm (e.g., gradient descent) needs to be designed such that given any

observation π, a desired control input u can be computed. This provides a model-based *implicit policy*, as shown in Fig. 1.3b. When the cost $\mathcal{J}$ is non-convex, the control input u computed online may only be a local optimum. The computation of u is only initiated when π is obtained in the execution phase, which is different from an explicit policy that computes u for all possible π in the design phase.

3. Model-free policy.
   We can also approximate the policy using parametric functions (e.g., neural networks) in the training phase. Given model-free knowledge which consists of a set of data points (π, u), the function g can be approximated from the data as shown in Fig. 1.3c. Since no explicit cost or model is extracted from the data, this is a model-free logic.[7]

The three-component behavior system represents a variety of existing methods. We summarize the methods into the following three categories ranging from nature-oriented design to nurture-oriented design, as shown in Fig. 1.3d.

1. Category 1: Pure Nature
   The designer specifies the cost and the model and designs model-based logics to solve the optimization without any learning process. Representative methods are (1) classic control methods and Markov decision processes (MDPs) where explicit policies are obtained in the design phase (e.g., in the control of flexible robot joints [100, 214], or in safety critical situations [80, 228]), and (2) model predictive control (MPC) methods where the optimization is computed in the execution phase [146, 46, 122].

2. Category 2: Nature + Nurture
   In one case, the designer specifies the cost, designs the logic explicitly, and defers to the learning process to identify the models. Classic adaptive control and adaptive MPC belong to this category. Application of this approach in human-robot interactions can be found in [185, 131, 76, 152]. The advantage of this method is that it can account for time varying environmental changes, especially when humans are in the loop, while the designer still has control over the task performance through explicit design of the knowledge and the logic.
   Another case is that the designer only designs the logic and the learning process explicitly. The knowledge is obtained in the training phase by either trial-and-error or expert demonstration. Representative methods are model-based reinforcement learning and inverse reinforcement learning such as apprentice learning [8, 15, 70]. The learning object can either be the cost

---

[7]Though a model-free policy can well approximate the true policy in data-rich regions, it may not generalize to regions with few data. Moreover, it is hard for model-free methods to infer hard constraints from the data. In Fig. 1.3c, the model-free policy (thick curve) deviates from the exact policy in regions with few data and even violates the hard constraint (shaded infeasible set).

(if the task is very complicated) or the model (if the dynamics are complicated) or both (as in cost-model coupled Q-learning). Application of this
approach in human-robot interactions can be found in [211, 11, 162]. The
advantage of this method is that mathematical modeling of the task and the
environment is no longer required in the design phase.

3. Category 3: Pure Nurture
   The designer designs the learning process explicitly, obtains model-free
   knowledge in the training phase, and approximates the logic using a model-
   free policy, which is usually represented by a neural network. In contrast
   to knowledge learning in Category 2, the knowledge in this category is not
   explicitly learned, but implicitly encoded in the data points and in the neural network. Representative methods are model-free reinforcement learning
   such as deep reinforcement learning [156] and imitation learning [104]. In
   addition to human subjects, the object to be imitated may be a behavior system in Category 1 and Category 2 [200]. This method is good for problems
   where (1) the task and the environment are extremely hard to model, (2) the
   state space is too large and (3) real-time computation is critical.

This classification may not be exhaustive. This book considers methods in Category 2 and Category 3. The existing methods in designing the three components will
be reviewed in the following sections.

1.3.2   KNOWLEDGE

As illustrated in Fig. 1.2a, knowledge may contain both raw data (e.g., measurement-
action pairs (π, u)) and the system principles extracted from raw data (e.g., the cost
$\mathcal{J}$ and the model $\mathcal{M}$). The former is also called model-free knowledge, while the latter is called model-based knowledge. For model-free knowledge, the measurement
in the raw data may contain (1) system states and (2) the instant cost evaluation on
current states and actions. In the following discussion, we call (π, u) the state-action-
cost pair, since it is a collection of system states, the corresponding desired actions
demonstrated by human or calculated by a solver, and the instant cost evaluated on
current states and actions. For model-based knowledge, the cost $\mathcal{J}$ provides incentives for the robot to finish the desired task efficiently as well as to interact safely
with other agents in the system, while the model $\mathcal{M}$ should be designed in order to
match the ground truth behavior of the environment. The cost $\mathcal{J}$ is a function that
assigns values to all measurement-action pairs, while the instant cost in raw data is
just a "point-wise" run-time measurement.

State-action-cost pairs

"All models are wrong, but some are useful."

George E. P. Box

Because of the complexity of the real world, a manual-designed model is usually a subjective simplification of the actual system. On the other hand, raw data is an objective observation of the world, which usually contains more comprehensive information than a simplified extracted model. Recently, many data mining methods have been developed, trying to recover useful information from raw measurement data. Naturally, the first and critical step for data mining is to construct a database as comprehensive as possible to fully embed the system property.

In robotics, the state-action-cost pairs provide a straightforward encoding of cluttered data. These pairs describe the evolutional system dynamics, the desired policy principle and the hidden cost mechanism implicitly. The state variables are the smallest possible subset of system variables that can represent the entire system condition at any given time. The action is the control command that can be applied on the system to influence the evolution of the system. The instant cost is a quantitative description of the goodness or badness of the current state and action. For a rational agent, it is supposed to be an optimizer which tends to achieve the highest rewards or the lowest costs by taking appropriate actions.

These data, i.e., the state-action-cost pairs, can be acquired by human demonstration or by robot self-exploration. In the demonstration approach, a human operator demonstrates the desired action w.r.t. a specific state, and labels the current condition with a corresponding cost value. In the exploration approach, the robot autonomously explores the state space by following some heuristic rules. The former approach collects more biased samples which contain human intention of optimality. This sampling bias might concentrate the learning in a limited region and boost the learning convergence. The latter approach, on the other hand, involves more randomness and explores broader regions. It provides more implementation flexibility since no human demonstration is required.

Cost

Cost $\mathcal{J}$ is a function on the state and action of the robot, which encodes a form of internal incentives for the robot to finish certain tasks. A rational robot always seeks to minimize the cost. The cost can either be a *static cost* or a *dynamic cost*. Static cost is a function on current state and action, which can be measured as the instant cost in the raw data discussed earlier. Dynamic cost is a function on a trajectory, i.e., a sequence of future states and actions, which is a discounted sum of all static costs along the trajectory.[8] Static cost includes runtime cost (e.g., the cost incurred before the completion of the task) and terminal cost (e.g., the cost related to the completion of the task). If the process never ends, there can be no terminal cost. For example, there is no terminal cost during lane following for automated vehicles. Dynamic cost can be summed either for a fixed time horizon (finite time or infinite time) or for a fixed task cycle. Fixed time horizon is adopted in linear quadratic regulator, preview control and receding horizon control. Fixed task cycle optimization is also called "minimum time problem" as the terminal time is treated as a decision variable. Fixed task cycle optimization is usually adopted in trajectory planning with fixed targets. The cost $\mathcal{J}$ will refer to the dynamic cost in the following discussion.

For model-based methods, the cost can either be designed or learned, with different levels of details. In Category 1, the cost is explicitly specified by a designer to achieve certain design objectives. However, for complicated tasks, it is hard for a human designer to construct desired cost functions, while it is easier to let the robot learn from experience or learn from others. In Category 2, the cost is obtained from data, i.e., the state-action-cost pairs discussed above. In particular, model-based reinforcement learning allows the robots to learn static cost from past experience, which can then be combined to formulate dynamic cost. Inverse reinforcement learning allows the robot to infer the cost $\mathcal{J}$ from some observed policy $\hat{g}(\cdot|\mathcal{J}, \mathcal{M})$, e.g., from an expert. The design of $\mathcal{J}$ will be discussed in the following chapters w.r.t. specific applications.

Model

Since the robot model is usually known in advance, we focus on the models of other intelligent entities in the environment, especially humans. As pointed out in [43], the biggest challenge in human-robot collaboration comes from human factors. To make co-robots human-friendly, human behavior needs to be modeled, learned, and predicted [14, 32, 176, 230]. The following three kinds of models are frequently used to describe human behaviors.

1. Reactive model.
   Under a reactive model, human dynamics are described using ordinary differential or difference equations. The inputs to the model are the identified features that affect human behaviors [191], while the output is the human motion.

2. Rational model.
   Under a rational model, a human is treated as a cost minimizer [160, 217], whose cost function depends on both his or her behavior and the robot's behavior. The human will reason his or her best action given his or her belief in the robot's behavior.

---

[8]In the MDP context, the static cost is equivalent to the reward function and the dynamic cost is equivalent to the value function or the Q function.

3. Bayesian model.

Under a Bayesian model, a human is treated as a stochastic agent [215]. A Bayesian network [170, 179] is associated with his or her decisions, i.e., the human's motion follows a probabilistic distribution conditioned on the robot's behavior.

It is usually hard to obtain the cost function or probability distribution that precisely describes human behavior in sophisticated problems. Moreover, it is dangerous to assume that a human will always adapt his or her motion to a robot's motion, as it intrinsically implies that the robot can "control" the human [20]. Since no universal model exists that can precisely describe human behavior, learning of human behaviors is necessary.

1.3.3 LOGIC

As discussed in section 1.3.1 and illustrated in Fig. 1.3, there are two kinds of logic, i.e., model-based logic and model-free logic. The model-based logic assumes that the system follows some hidden deterministic rules. One deterministic rule can be that agents are rational. Under the rationality assumption, the control policy is designed to be an optimizer which explicitly calculates optimal actions for the robot to achieve optimal performance. The optimal performance is described by minimal cost in this book. The cost metric as a part of knowledge needs delicate design as discussed in section 1.3.2. In contrast, the model-free logic regards the system as a stochastic process, where the control policy is nothing but a regression function between inputs (states) and outputs (actions). This regression function can be approximated purely by data mining, or modeled partly by system analysis. This section discusses methods to design model-based logic. Learning of model-free logic will be discussed in section 1.3.4.

The fundamental problem of a model-based logic is to find the minimum in the cost $\mathcal{J}$, given the data π and the model $\mathcal{M}$. In engineering applications, as the problems are complicated and in large scale, minimum seeking is non-trivial and various levels of approximations are needed. A model-based explicit policy is suitable for well-defined tasks and environments, however, hard to cover a variety of complicated scenarios that co-robots may face. This book focuses on model-based implicit policies which require online optimization. As we focus on physical movements in robot behaviors, the behavior design problem is essentially a motion planning problem. Depending on the look-ahead horizon of the dynamic cost, there are two kinds of motion planners, i.e., global planner (or long-term planner) and local planner (or short-term planner).

Global planner

A global planner evaluates the dynamic cost in a relatively long time horizon. The horizon may go to infinity. A comprehensive model of the environment is needed.

The existing global motion planning methods fall into two categories: planning by construction and planning by modification. Planning by construction refers to methods that extend a trajectory by attaching new points to it until the target point is reached. Search-based methods such as A* or D* search [195] and sampling-based methods such as rapidly-exploring random tree [105, 108] are typical planning-by-construction methods. Planning by modification refers to methods that perturb an existing trajectory so that the desired property is obtained. Optimization-based motion planning methods belong to this category. Gradient descent performs the perturbation or modification of the trajectory when solving the optimization [90, 189, 56]. Every type of methods has pros and cons. The trajectories planned by construction are more feasible than trajectories planned by modification. However, as the space is usually discretized during trajectory construction, the constructed trajectories are not as smooth as the trajectories planned by modification. This book focuses on optimization-based methods. If the environment is perfectly known without any uncertainty in advance, the global planner is capable of finding the global optimal trajectory. However, in practice, it is computationally expensive to build the model of the environment given the limited sensing ability and computational resource.

Local planner

A local planner tries to only plan a few steps ahead based on the limited knowledge of the environment obtained by current sensory data. Such planners require less computation power, and are good choices for robots with limited sensing abilities. An extreme case of a local planner is the reactive controller where the planning horizon reduces to one. There are usually closed-form solutions for reactive controllers. A closed-form solution of reactive control is a direct mapping from the sensory data to the control, which can greatly relieve the computation burden.[9] Existing reactive methods include potential field method [95], barrier certificate [12], sliding mode method [74], and a family of biologically inspired methods [23]. However, as it only regulates the motion locally, a local planner is sensitive to local optima. It is possible for the robot to get stuck in some location and cannot reach the target. Thus, the global convergence to the target needs to be addressed when designing a local planner. In general, it is very hard to guarantee global convergence for local planners, as the system under consideration is nonlinear (as the robot motion is nonholonomic), time-varying (as the obstacles in the system are moving and deforming), and stochastic (as the information about the environment is limited).

To leverage the advantages of different planners, a global planner and a local planner can be combined in parallel, which will be discussed in Chapter 2.

---

[9]A reactive control policy is similar to an explicit policy in Fig. 1.3a in that they are both direct mapping from the sensory data to the control. However, only local information is considered in the reactive control policy, while global information is encoded in the explicit policy when it is computed in the design phase.

1.3.4   LEARNING PROCESS

Learning process updates the knowledge or the logic according to the observed data. The ultimate goal of learning is to derive a better logic from the updates. As mentioned in section 1.3.1, there are model-based and model-free logics. In model-based approaches, learning process directly updates the knowledge, so that the logic can be updated based on physical principles and deductions [16]. In model-free approaches, the learning process directly updates the logic.

Updating knowledge

In a model-based approach, both the cost and the model can be updated. This book considers the learning on the model, which is a *cognition skill*. In particular, interpretation and prediction of human motions are considered. Learning on the model can be done both offline (in the training phase) and online (in the execution phase). Offline learning is usually used to classify human behaviors and to identify models that describe human behaviors as discussed in [19, 64, 178]. On the other hand, online learning is capable of adapting the offline learned models to humans' time-varying behaviors online as discussed in [72, 185]. Methods to learn human model will be discussed in Chapter 5.

Updating logic

In model-based approaches, the logic is indirectly updated when the knowledge is updated. Such an approach is good for system transparency and interpretability, which is critical in many application domains, especially for industrial applications.

However, as robots are deployed to more sophisticated tasks, such as grasping dexterous objects and manipulating deformable objects, a pure model-based learning approach runs into its bottleneck. In these scenarios, either the system complexity goes beyond our capability to model, or the existing models cannot introduce satisfying results. On the other hand, along with the development of machine learning methods, especially the breakthrough of deep learning in recent years [110], people realize that: although human cognition process is complicated to model, it could be alternatively regressed by high dimensional parametric functions given a large amount of training data [157]. With this observation, many model-free learning approaches have been proposed recently, trying to teach robots new skills from data, instead of modeling those skills from principles. In the training stage, thousands or even millions of demonstration data is collected either by offline expert demonstration (imitation learning [24, 8], learning from demonstration [15, 17]), or by online trial and error (model-free reinforcement learning [155, 194]). The collected samples are then utilized to fit an optimal control policy g such that $u = g(\pi)$. These model-free learning approaches have universal structure and common training process, which can be quickly deployed to various scenarios. However, many of them are working in a black box manner. As a consequence, it can be hard to interpret the outcomes of model-free learning. The stability of the learning system is also difficult to evaluate. Therefore, model-free learning are usually deployed for high-level tasks (e.g., vision detection), while the low-level implementation (e.g., motor servoing) is still based on model-based learning approaches. This hierarchical structure achieves a balance between performance and stability of the overall system. It is notable that for non-safety-critical scenarios, end-to-end learning [113, 31] is favored, where the hierarchical structures of systems are embedded in a single framework of deep neural networks, and thus the maximum potential of neural networks is exploited.

The model-free learning approaches, together with the model-based learning approaches which need online system identification, are called data-driven approaches in general. The learning methods will be discussed in Chapters 5 and 6.

1.4   DESIGN OBJECTIVES

Intelligent robots are multi-purposed complex entities. Their behavior systems (introduced earlier in section 1.3) need to be designed such that the robots are safe during human-robot interactions, efficient in task performance, able to imitate behaviors of other intelligent entities, dexterous to adapt to different environments and tasks, and cooperative with other intelligent entities. This book studies methodologies to achieve these five design objectives in Part II. The challenges and existing solutions to address these design objectives are discussed below.

1.4.1   SAFETY

Human safety is one of the biggest concerns in HRI [204]. Robots should be both intrinsically safe, e.g., minimizing impact if collision happens [79], and interactively safe, e.g., avoiding collision at the first place [216]. Interactive safety needs to be addressed in the context of decision making, motion planning and control of the robot during HRI.

The difficulty of designing safe robot behavior comes from (1) sophistication of human motion and (2) real-time implementation of the algorithm. Control theory offers plenty of methods to deal with disturbances. However, when humans are in the loop, the disturbances introduced by human motions are much more complicated. They are time varying with unknown models. Moreover, it seems counterproductive to view human motions as disturbances in the robot control, as the system should be human-oriented. Learning of human behaviors is needed to make robots compliant. Even with robots learning human behaviors, another concern is that a complex control algorithm may not be executed fast enough to ensure quick responses of robots in emergencies.

Conventional approach to address the interactive safety is conservative, which slows down the robot when humans are nearby, hence sacrifices productivity for the sake of safety. Less conservative methods in the context of obstacle avoidance have also been used to address safety in HRI, e.g., potential field methods [168, 95], sliding mode methods [74]. These two methods result in closed-form analytical control laws, but are not optimal. Some authors formulate the problem as an optimization or optimal control problem with hard constraints to represent the safety requirements [56]. Unfortunately, these non-convex optimization problems are generally hard to solve analytically [86]. Different approximations and numerical methods are used to obtain solutions [189, 195]. However, these numerical methods may not be executed fast enough for timely responses in emergency situations.

Chapter 3 discusses the approaches that we developed for safe human-robot interaction. The problem is formulated as an optimization with hard safety constraints, which is non-convex. We then transform the non-convex state space constraint into a convex control space constraint using the idea of invariant set. In this way, the problem is transformed to a convex problem which can be solved efficiently online. This method can also take the uncertainty in the human behavior into account. It has been shown that the safety constraint will never be violated using the safe control method if the human behavior satisfies certain assumptions.

1.4.2 EFFICIENCY

Robot efficiency is essential to maximize task performance. Robots should understand the current situation, then plan and execute the proper actions to complete the task in minimum time. In human-involved dynamic environments, robot motion needs frequent replanning to cope with time-varying tasks. Hence, computation efficiency of the motion planning algorithms is critical to ensure real-time response of the robot.

The challenge of efficient motion planning results from the complexity of the planning problem, which may be highly nonlinear due to the dynamic constraints, and highly non-convex due to the constraints for obstacle avoidance. It is hard to solve the planning problem in real time.

This book focuses on optimization-based motion planning. There are various methods developed to deal with the non-linearity and non-convexity [163, 199] in optimization-based motion planning. One popular way is through convexification [210], e.g., transforming the non-convex problem into a convex one. Some authors tried to transform the non-convex problem to semidefinite programming [58]. Some authors proposed to introduce lossless convexification by augmenting the space [9, 83], and some authors proposed successive linear approximation to remove non-convex constraints [141, 142]. However, the first method requires the cost function to be quadratic. The second approach depends highly on the linearity of the system and may not be able to handle various obstacles. Finally, the third approach may not generalize to non-differentiable problems. One of the most popular convexification methods is the sequential quadratic programming (SQP) [198, 213], which approximates the non-convex problem as a sequence of quadratic programming problems and solves them iteratively. The method has been successfully applied to offline robot motion planning [91, 189]. However, as SQP is a generic algorithm, the unique structure of the motion planning problems is neglected, which usually results in failure to meet the real-time requirement in engineering applications.

Chapter 4 discusses the approaches that we developed to ensure efficient robot motion planning during human-robot interactions. In practice, the cost function for motion planning is designed to be convex [221, 177], while the non-convexity mainly comes from the physical constraints, e.g., robot dynamics and collision avoidance. By exploiting the geometric structure of the problems, we propose the convex feasible set algorithm [127] and the slack convex feasible set algorithm [136] to handle motion planning problems with convex objective functions and non-convex constraints. It will be shown that these methods perform better than generic methods in robot motion planning in terms of efficiency.

1.4.3 IMITATION

Imitation is the ability of robots to reproduce human-like movements. On the one hand, by learning to mimic human behavior, robots can better predict human's future motion through mental simulation. Better prediction of human motion can lead to smoother human-robot interaction. On the other hand, robots can physically behave more like humans, so as to accomplish dexterous tasks which usually require human skills.

Take robotic precision assembly as an example. Tuning an appropriate force controller for a robot is usually a non-trivial task. In contrast, humans can accomplish assembly tasks compliantly with much less time and fewer trials. Human's inherent assembly skill is achieved by a form of internal control, which has been well tuned through our daily operation and experience. It is more intuitive and efficient to teach robots insertion skill by mimicking human's compliant motions, compared to designing a new controller from scratch [206, 114, 44, 101].

In general, the challenges for robots to mimic human behavior come from two aspects. First, it is difficult to select the proper variables to sufficiently and efficiently depict human behavior, while those variables should be observable and measurable by available sensors. Second, it is difficult to properly encode the measured human behaviors, and to reproduce human-like motions in novel scenarios. There are specific challenges associated with imitation for prediction [201, 224] and imitation for action [39, 40, 187]. Regarding imitation for prediction, the major problem is homogeneity of the learned models, which ignores individual differences. Hence, to make accurate "personalized" predictions, online adaptation of those models are necessary. Regarding imitation for action (or skill learning), few of those existing methods are implementable on industrial robots. Unlike collaborative robots, industrial robots are highly rigid because of large gear reduction ratio on the drive train. For assembly tasks, it implies that there is no compliance on the industrial robot's mechanism to assist assembly. Besides, the clearance requirement for industrial assembly is much more strict. Therefore, new schemes which are implementable on industrial robots need to be developed to support imitation learning.

Chapter 5 discusses methods to (1) learn individualized human behavior models so as to make better predictions; and (2) teach industrial robots assembly skills from human demonstration with varying admittance. For prediction, we introduce online learning. For skill learning, we introduce a model-free framework with guaranteed stability from Lyapunov analysis.

1.4.4   DEXTERITY

Dexterity is the ability of robots to change its nominal motion patterns to adapt to the change of its surrounding environment or task requirements.

Although there are numerous approaches for robot learning, most of them follow a similar pipeline: learning the logic in the training stage, and then applying the learned logic in new scenarios. It is implicitly assumed that the learned logic is universally applicable. However, it remains challenging to verify that the learned logic is truly universal and dexterous. To make robots dexterous, it will be beneficial to first understand the dexterity of human and then equip robots with the same ability.

Many studies in cognitive science shows that humans can make dexterous decisions w.r.t. various scenarios through analogical reasoning [42, 164, 116]. Given a new scenario, humans will try to find its correlation with past scenarios, and transfer previous problem-solving actions to the new situation. With this analogy approach, the problem can be solved without understanding the universal policy or causality behind the system. This analogy approach also enables humans to learn efficiently from small data samples.

Chapter 6 discusses an analogy-based learning approach to make robots dexterous. Unlike traditional model-based or model-free approaches, which try to formulate control policies, this new approach focuses on finding correlations among scenarios. This approach has two stages: the reminding stage and the transferring stage. During the reminding stage, the similarity between the past scenarios and the current scenario will be calculated. The most similar pair will be selected and a mapping function between the two scenarios will be constructed by non-rigid registration. During the transferring stage, the past actions will be transformed by the mapping function to generate new actions in the current situation. This approach is plausible for applications which are hard to model or to collect data from. More properties and advantages of analogy-based learning will be introduced in details in Chapter 6.

1.4.5   COOPERATION

Cooperation is the ability of robots to achieve a mutual goal together with other intelligent entities (e.g., humans or other robots) without centralized coordinations.

Cooperation is hard to achieve mainly due to information asymmetry [175], i.e., the parameter in the behavior system of one agent is unknown to another agent. Information asymmetry is common in HRI. From the agent perspective, most learning and control strategies follow a centralized design philosophy, which adjusts the agent's estimation if its observation deviates from its prediction. This strategy is suitable in the single agent case, since the deviation can only be caused by the agent's wrong estimation. However, in a multi-agent system, where each agent acts on its estimation of the unknowns, the gap between an agent's observation and prediction is not only caused by the agent's wrong estimation (wrong prediction), but also other agents' wrong estimations and subsequent improper actions (deviated observations). Interactions and other agents' sub-optimality need to be considered explicitly in agent behavior design.

Conventionally, to compensate information asymmetry during online execution, distributed control laws are designed for each robot agent [180]. Following the control laws specified by the intelligent designer, who has an omniscient perspective, the agents do not need much intelligence. Although it is not explicitly stated that the control laws are globally known, the effectiveness of the controllers is based on the confidence that the designed control laws are perfectly followed by all agents. In this case, an agent just needs to follow its own control law, while other agents will take care of everything else by following their designed control laws. This strategy usually works well on problems with predefined environment and interaction patterns, but has limited extendability, as it is hard to design a universal control law that leads to desired results in a wide variety of situations, especially when the system topology is time-varying (interactions are happening among different agents at different times).

Chapter 7 discusses two ways to encourage cooperation. The first method introduces compensations of other agents' wrong estimations in the ego agent's learning strategy. In this way, even without explicit communication, the multi-agent system can still achieve the mutual goal. The second method allows agents to communicate with each other to resolve the conflicts. A conflict resolution mechanism is developed to form a consensus.

1.5   SYSTEM EVALUATION

The designed robot behaviors need to be evaluated in human-robot systems. The evaluation can be performed theoretically as well as experimentally.

1.5.1   THEORETICAL EVALUATION

During theoretical analysis, the questions to answer are:

1. Is the logic optimal and safe?
2. Will the learning process generate converging sequences to the ground truth?
3. Will the designed behaviors lead to desired outcomes of the multi-agent system?

The first two questions are modular-wise. The third question is system-wise which concerns with the stability, robustness stability and optimality of the closed loop system, e.g., whether the closed loop multi-agent system is self-organized [140]. System level analysis is challenging due to (1) the complication of interactions among different agents, (2) the difficulty in justifying the assumptions on human behavior; (3) the insufficiency of existing tools in game theory to analyze suboptimal agents.[10]

1.5.2 EXPERIMENTAL EVALUATION

In addition to theoretical analysis, experiments should be conducted to evaluate the performance of the designed robot behaviors. However, to protect human subjects, it is desirable if we can separate human subjects and the robot physically during the early phase of deployment. Figure 1.4 shows five different platforms that have been developed for experimental evaluations.

The first platform is the virtual reality-based human-in-the-loop platform. The robot's motion is simulated in a robot simulator. A human subject observes the robot movement through the virtual reality display (e.g., virtual reality glasses, augmented reality glasses or monitors). The reaction of the human subject is captured by sensors (e.g., Kinect or touchpad). The sensor data is then sent to the robot behavior system to compute the desired control input. The advantage of this platform is that it is safe to human subjects and convenient for idea testing. The disadvantage is that the robot simulator may neglect dynamic details of the physical robot, hence reducing the reliability of the results.

The second platform is the virtual reality-based hardware-in-the-loop platform. Its difference from the first platform is that the robot motion is no longer simulated, but directly measured from the robot hardware. This solves the problem of dynamic mis-match that the robot simulator may have. However, as the interaction happens virtually, the human subjects may not react in the same way that they do with real robots.

The third platform is the dummy-robot interaction platform. The interaction happens physically between the dummy and the robot. The human directly observes the robot motion and controls the dummy to interact with the robot. As there are physical interactions, an emergency check module should be added to ensure safety. The advantage of this platform is that it is safe to human subjects, while it is able to test interactions physically. However, the disadvantage is that the dummy usually has fewer degrees of freedom than a human subject.

The fourth platform is the robot-robot interaction platform. Each robot regards the other as a "human". In general, such a platform allows us to inquire whether the designed behavior system can cope with intelligent entities other than humans. Moreover, it can examine whether the multi-agent system consisting of those intelligent robots are stable or not.

---

[10]The basic assumption in game theory is that all agents are rational, e.g., always behaving optimally. However, the logic that we designed for the robot may not be optimal.

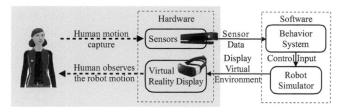

(a) Virtual reality-based human-in-the-loop platform.

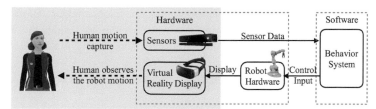

(b) Virtual reality-based hardware-in-the-loop platform.

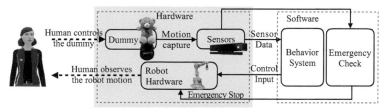

(c) Dummy-robot interaction platform.

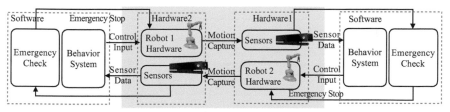

(d) Robot-robot interaction platform.

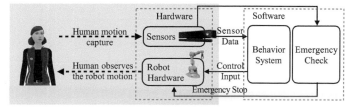

(e) Human-robot interaction platform.

Figure 1.4: The evaluation platforms.

The fifth platform is the human-robot interaction platform where human subjects directly interact with robots. Such platform should be designed to illustrate the performance of the robot in real tasks.

In Part III of the book, evaluation results on those platforms will be discussed.

1.6   OUTLINE OF THE BOOK

This book aims to establish a set of methodologies for robot behavior design in order to maximize their performance in human-involved dynamic uncertain environments. The emphases of design objectives are safety, efficiency, imitation, dexterity, and cooperation.

As illustrated in Fig. 1.5, the remainder of the book is organized as follows.

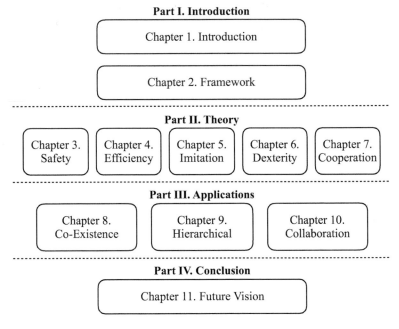

Figure 1.5: Overview of the book content.

Part I is for introduction. Chapter 2 overviews the mathematical methods for robot behavior design in a multi-agent framework.

Part II discusses theory. Chapter 3 is for safety. Chapter 4 is for efficiency. Chapter 5 is for imitation. Chapter 6 is for dexterity. Chapter 7 is for cooperation.

Part III discusses applications. Chapter 8 discusses space-sharing human-robot interactions, such as the interaction between industrial co-robots and human workers, and the interaction between autonomous vehicles and other road participants. Chapter 9 discusses hierarchical human-robot interactions where a robot learns skills from human teachers. Chapter 10 discusses time-sharing human-robot interactions where a robot and a human collaborate to finish a joint task.

Part IV discusses future vision and concludes the book.

Each chapter is intended to be self-contained. Some of the work has been published in [117, 118, 119, 120, 124, 127, 128, 130, 131, 132, 133, 134, 135, 136, 137, 138, 139, 206, 207, 208, 209].

2 Framework

Human-robot interactions can be modeled in a multi-agent framework, where robots and humans are regarded as agents. In this chapter, a multi-agent model will be introduced first, followed by discussions of agent behavior design and architecture.

2.1 MULTI-AGENT FRAMEWORK

2.1.1 THE GENERAL MULTI-AGENT MODEL

An agent is defined as an independent autonomous entity which has the "see, think and do" ability, i.e., an agent observes through sensors, acts upon the environment using actuators, and directs its activity towards achieving goals. The identification of individual agents depends on scenarios. For interactions among human workers and industrial robots, a human worker can be regarded as an agent. For interactions among automated vehicles and human-driven vehicles, a human driver together with the vehicle are viewed as one agent. Moreover, if a group of robots are coordinated by one central decision maker, they may be regarded as one agent.

Suppose there are N agents in the environment and are indexed from 1 to N. Denote agent i's state as x_i, its control input as u_i, its dataset as π_i for $i = 1, ..., N$. The physical interpretations of the state, input and dataset for different plants in different scenarios vary. For a robot arm, the state can be joint position and joint velocity, while the input can be joint torque. For an automated vehicle, the state can be vehicle position and heading, while the input can be throttle and pedal angles. When communication is considered, the state may also include the transmitted information and the input can be the action to send information. Let x_e be the state of the environment.

Denote the system state as $x = [x_1^\mathsf{T}, \ldots, x_N^\mathsf{T}, x_e^\mathsf{T}]^\mathsf{T} \in \mathcal{X}$ where $\mathcal{X}$ is the state space of the system. The open loop system dynamics can be written as

$$\dot{x} = f(x, u_1, u_2, \ldots, u_N, w), \tag{2.1}$$

where f is the system dynamic function and w is the dynamic noise.

According to the discussion in section 1.3 and Fig. 1.2a, agent i's behavior system generates the control input u_i based on the dataset π_i,

$$u_i = \mathcal{B}_i(\pi_i), \tag{2.2}$$

where the function $\mathcal{B}_i$ contains the three components (knowledge, logic and learning). When there is no learning process, then $\mathcal{B}_i(\pi_i) = g_i(\pi_i | \mathcal{J}_i, \mathcal{M}_i)$ where g_i is the logic function of agent i and $\mathcal{J}_i$ is the cost function of agent i. Agent i's model $\mathcal{M}_i$ of the environment includes the estimates of the system dynamics (2.1) and other

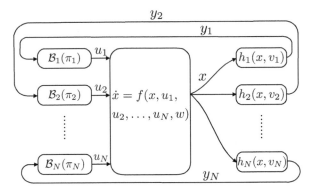

Figure 2.1: Block diagram of a multi-agent system.

agents' behaviors, e.g., the function $\mathcal{B}_j$ for $j \neq i$. The dataset contains observations on the system state x, which is a combination of the measured data and the communicated information. Agent i's dataset at time T contains all the observations $y_i(t)$ from the start time t_0 up to time T, i.e., $\pi_i(T) = \{y_i(t)\}_{t \in [t_0, T]}$ where

$$y_i = h_i(x, v_i),\qquad(2.3)$$

and v_i is the measurement noise.

Applying (2.2) and (2.3) in the open-loop dynamics (2.1), the closed-loop dynamics become

$$\dot{x} = \mathcal{F}(x, w, v_1, ..., v_N),\qquad(2.4)$$

where the closed-loop dynamic function $\mathcal{F}$ depends on the open-loop dynamic function f, the measurement functions h_i for all i, and the behavior systems $\mathcal{B}_i$ for all i. The system block diagram for the general multi-agent system is shown in Fig. 2.1 according to (2.1), (2.2) and (2.3).

2.1.2   MODELS FOR DIFFERENT MODES OF INTERACTIONS

As discussed in section 1.2, the relationship between humans and robots is complicated. Nonetheless, the general model in Fig. 2.1 can cover the diversity of HRI. In the following discussion, we show how different scenarios can fit into the general model. In particular, the model can describe systems no matter whether time synchronization among agents is required or not, no matter complete information is available or not, and no matter whether agents make decisions sequentially or simultaneously.

Space-Sharing Interactions and Time-Sharing Interactions

In the general model discussed in section 2.1.1, system dynamics is described in a coupled form in the open loop. Since every agent has its own dynamics, we may

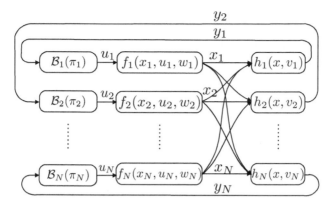

Figure 2.2: Block diagram of a decomposable multi-agent system.

decouple the open-loop system. For agent i, its dynamics in the open loop can be described as

$$\dot{x}_i = f_i(x_i, u_i, w_i, e_i), \tag{2.5}$$

where e_i represents external inputs induced by the environment or other agents. When there is no direct contact among agents, $e_i = \emptyset$. Otherwise, $e_i \neq \emptyset$ and it comes in pairs, i.e., if agent i receives an external input from agent j, then agent j also receives an external input from agent i. The external input pair can be viewed as a constraint between two agents, such as the force and reaction between a human patient and an exoskeleton. When there is no constraint among agents, the system does not need to be synchronized in time. Such interaction is called space-sharing interaction [102]. When there are constraints among agents, the system needs to be synchronized and it is called time-sharing interaction. The block diagram for the space-sharing system is shown in Fig. 2.2.

Complete Information and Incomplete Information

When there is no learning process, agent i's control input u_i is chosen by minimizing its cost function $\mathcal{J}_i(x, u_1, \cdots, u_N)$, given π_i and $\mathcal{M}_i$, i.e.,

$$u_i = g_i(\pi_i \mid \mathcal{J}_i, \mathcal{M}_i) = \arg\min_{u_i} \mathbb{E}[\mathcal{J}_i(x, u_1, \cdots, u_N) \mid \pi_i, \mathcal{M}_i], \tag{2.6}$$

where the cost function depends on the system state x and other agents' inputs, and on the dataset π_i and agent i's model $\mathcal{M}_i$ of the environment.

If the system dynamics and agents' costs are globally known, and all agents' models of the environment are correct, the system is with *complete information*. If any of the information is unknown to some agents, e.g., some agent's model of the environment does not match the ground truth of the environment, the system is with *incomplete information*. If some agent knows information that others don't know, e.g., an agent knows its own cost while others don't, it is called *information asymmetry*. Information asymmetry is common in human-robot systems as it is difficult to obtain a human's internal cost function in advance.

Sequential System and Simultaneous System

As discussed in section 1.2, human-robot relationships can either be *parallel* or *hierarchical*, depending on how the agents' decisions are made. The decision hierarchy is related to the information structure of the system [22], e.g., what data is available to which agent when the agents are making decisions in (2.6).

For a parallel relationship, no agent can obtain more data before they make the next move. This is called a *simultaneous game*. For a hierarchical relationship, some agent obtains more data about others before they take the next move. The agents with fewer data are considered as leaders and the agents with more data are considered as followers. For example, in robot skill learning from human demonstration, the human teacher is a leader and the robot learner is a follower. The robot learner only moves after collecting enough human demonstration data, while the human teacher teaches the robot from the very beginning without any robot motion data. A leader is supposed to comprehensively evaluate the consequences of its action by taking into consideration the follower's reactions. On the other hand, a follower only needs to follow the leader's plan. This is called a *sequential game*.

2.1.3   FEATURES OF HUMAN-ROBOT SYSTEMS

Although HRI can be modeled in a multi-agent framework, it is different from conventional multi-agent models [225] both in the design phase and in the execution phase. The distinct features are summarized below:

1. Partial knowledge in the design phase:
   The system is only partially known even to the designer in the design phase. It is common for a robot in the execution phase to have partial knowledge since it is beneficial to scale down the problem. However, the successful implementation of those systems depends on the designer's understanding of the overall system in the design phase. As there is no universal model to describe human behavior in the presence of fully automated robots, the system contains many structural or hyper-parametric uncertainties in the design phase. Methods to deal with those uncertainties need to be studied.

2. Time-varying topology in the execution phase:
   When a robot is brought to life in the execution phase, it may encounter various agents in diverse scenarios. For example, automated vehicles need to interact with different road participants in various traffic conditions, which implies that different agents will be added or removed in the system model in Fig. 2.1. The topology of the multi-agent system is time-varying. In comparison, conventional multi-agent networks such as power networks usu-

ally have relatively static topology. The time-varying topology brings large structural uncertainty to the design and analysis of the multi-agent systems.

Regarding the features mentioned above, we are trying to solve the following problems given the framework in Fig. 2.1. The first problem is how to model the physical world under the multi-agent framework. For example, what is the proper dynamic model f? What is the behavior model $\mathcal{B}_i$ for human agent i? How are different skills represented in the knowledge module? The second problem is how to design the behavior system $\mathcal{B}_i$ for a robot agent i given the model of the physical world. For example, how to ensure that the designed behavior is safe, efficient, dexterous, cooperative, and able to imitate other intelligent entities? The third problem is how to properly analyze and evaluate the performance of the designed behavior. For example, will the multi-agent system result in instability given the designed agent behavior? The three problems represent modeling, design, and verification aspects respectively, which will be answered in the following chapters.

In the remainder of this chapter, we discuss the architecture of the proposed behavior system for robot agents. Note that human behavior can also be tuned by providing different incentives to shape their internal costs and models. Readers are referred to [21, 10] for detailed discussions on how to provide desired incentives.

2.2 AGENT BEHAVIOR DESIGN AND ARCHITECTURE

Behavior design for robots deals with the three components discussed in section 1.3 and shown in Fig. 1.2a. The design can either be model-based or model-free, as discussed in section 1.3. In the following two sections, we overview a model-based design approach and a model-free design approach. Both approaches will be used in the following chapters.

2.2.1 MODEL-BASED DESIGN

We formulate the model-based design in the framework of adaptive model predictive control, which deals with the design of $\mathcal{J}$ and $\mathcal{M}$, the design of the logic g, and the design of the learning process to update $\mathcal{M}$.

Knowledge: The Optimization Problem

Model-based knowledge consists of cost and model. To provide correct incentives to the robot agent i in the multi-agent system, its cost is designed with the following structure,

$$\mathcal{J}_i(x, u_1, u_2, \cdots, u_N) = J_i(x, u_1, u_2, \cdots, u_N), \tag{2.7a}$$

$$s.t. \quad u_i \in \Omega, x_i \in \Gamma, \dot{x}_i = f_i(x_i, u_i, w_i, e_i), \tag{2.7b}$$

$$y_i = h_i(x, v_i), \tag{2.7c}$$

$$x \in \mathcal{I}, \tag{2.7d}$$

where the right hand side of (2.7a) is the dynamic cost for task performance. It is evaluated over a planning / look-ahead horizon T. The horizon T can either be chosen as a fixed number or as a decision variable that should be optimized up to the completion of the task. Equation (2.7b) is the constraint for the agent itself, i.e., constraint Ω on the control input (such as control saturations), constraint Γ on the state (such as joint limits for robot arms), and the dynamic constraint represented by an ordinary differential equation. Equation (2.7c) is the measurement constraint, which builds the relationship between the state x and the dataset π_i. Equation (2.7d) is the constraint induced by interactions in the multi-agent system, e.g., safety constraint for collision avoidance, where $\mathcal{I} \subset \mathcal{X}$ is a subset of the system's state space. Equation (2.7a) can also be called the soft constraint, while (2.7b) to (2.7d) can be called the hard constraints. The hard constraints (2.7b) to (2.7c) are determined by physical properties of the agent. The interaction constraint (2.7d) depends on the requirement of the multi-agent system. The detailed design of the cost will be discussed with respect to different design objectives in Part II, and with respect to different applications in Part III.

Agent i's model of the environment is an estimation of the system dynamics and other agents' behaviors, i.e., all blocks that are unknown to agent i in Fig. 2.1. However, both the dynamics of the system and other agents' behavior models are hard to obtain in the design phase, especially for human-robot systems. Hence, the model $\mathcal{M}_i$ will be identified in the learning process. Moreover, since only the closed-loop dynamics of other agents matter, we only need to identify a coupled model,

$$\dot{x}_{-i} = F_i(x_{-i}, x_i, v_{-i}, w_{-i}), \tag{2.8}$$

where the subscript $-i$ refers to the combination of variables except for agent i, i.e., $x_{-i} = [x_1^\mathsf{T}, \ldots, x_{i-1}^\mathsf{T}, x_{i+1}^\mathsf{T}, \cdots, x_N^\mathsf{T}, x_e^\mathsf{T}]^\mathsf{T}$. The function F_i depends on the open-loop dynamic function f and other agents' behavior models $\mathcal{B}_j$ for $j \neq i$. And $\mathcal{M}_i$ is an estimate of F_i, i.e., $\mathcal{M}_i = \hat{F}_i$. The model $\mathcal{M}_i$ is needed to estimate others' control u_j for $j \neq i$ in (2.7a) and to compute the system state x in (2.7d).

Model-Based Logic: The Motion Planning Skill

In the model-based design, a logic is needed to obtain the optimal action that minimizes the cost given the model as shown in (2.6). A "see-think-do" structure is adopted to tackle the complicated optimization problem (2.7) as shown in Fig. 2.3. The mission of the "see" step is to process the high dimensional data $\pi_i(t)$ to obtain an estimate of the system states $\hat{x}(t)$. The "think" step is to solve an approximated optimization problem using the estimated states and construct a realizable plan $x_i(t : t+T)$ in a time window of length T. The "do" step is to realize the plan by generating a control input $u_i(t)$.

In the "think" step, two important modules are *prediction* and *planning*. In the prediction module, agent i makes predictions of other agents $\hat{x}_{-i}(t : t+T)$ based on the model $\mathcal{M}_i$. The prediction can be a fixed trajectory or a function that depends on agent i's future state $x_i(t : t+T)$. When the prediction is a fixed trajectory, the others' reactions to the ego agent are not considered. In the planning module, the future

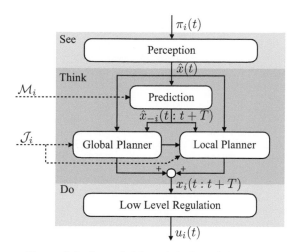

Figure 2.3: A model-based design of the logic.

movement $x_i(t:t+T)$ is computed given the current system state $\hat{x}(t)$, predictions of others' motion $\hat{x}_{-i}(t:t+T)$ as well as the task requirements and constraints encoded in $\mathcal{J}_i$.

As it is computationally expensive to obtain the optimal solution of (2.7) for all scenarios in the design phase, the optimization problem is computed in the execution phase given the obtained information in real time. This approach corresponds to the model-based implicit logic in Fig. 1.3b. However, there are two major challenges in real-time motion planning. The first challenge is the difficulty in planning a safe and efficient trajectory when there are large uncertainties in other agents' behaviors. As the uncertainty accumulates, solving the problem (2.7) in the long term might make the robot's motion very conservative. The second challenge is real-time computation with limited computation power since problem (2.7) is highly non-convex. We design a unique parallel planning and control architecture to address the first challenge (to be discussed below) and develop fast online optimization solvers to address the second challenge (to be discussed in Chapter 4).

A toy example is presented in Fig. 2.4 to illustrate the first challenge. There is a robot and a human in a closed environment. The robot is required to approach the target while avoiding the human. The time axis is introduced to illustrate the spatiotemporal trajectories. At the first planning step, the robot makes a prediction of the human trajectory and plans a trajectory for itself avoiding the uncertainty cone. As time propagates, the planned trajectory is executed and the human trajectory is observed. In the next planning step, the robot repeats the process: predicts the human trajectory, plans its own trajectory, and then executes the planned trajectory. The process is repeated in the following steps. This is the conventional model predictive control (MPC) method, which is safe, however, too conservative, as the robot hesitates to get close to the human due to the perceived uncertainty.

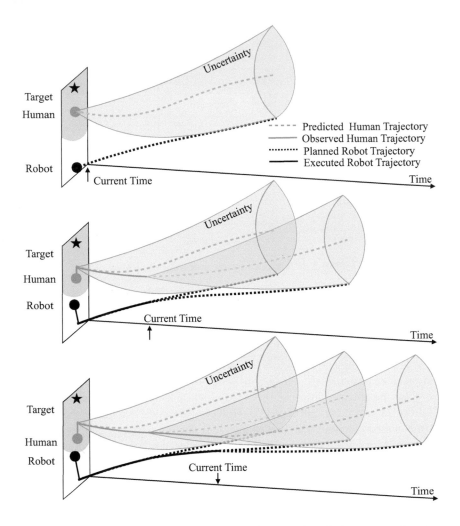

Figure 2.4: Illustration of the accumulation of uncertainty in long-term planning.

On the other hand, the uncertainty will not accumulate too much for a short term planner. However, using a short term planner alone will also be problematic. The robot can easily get stuck in local optima and not be able to finish the task as illustrated in Fig. 2.5. In this scenario, the robot does not have a global perspective on how to bypass the obstacle. Although it is possible to construct a globally-converging local policy for some simple dynamic systems when the environment satisfies certain geometric properties [186], it is in general hard to obtain a globally-converging local policy for robots with complicated dynamics in complicated environments.

A parallel planner which consists of a global (long-term) planner as well as a local (short-term) planner can be introduced to leverage the benefits of the two plan-

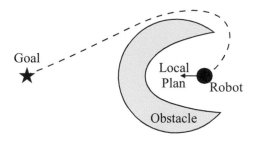

Figure 2.5: Illustration of the local optima problem with the local planner.

ners. The idea is to have a long-term planner solving (2.7) and considering only a rough estimation of human's future trajectory, and a short-term planner addressing uncertainties and enforcing the safety constraint (2.7d). The long-term planning is efficiency-oriented and is analogous to deliberate thinking. The short-term planning is safety-oriented and is analogous to a reflex behavior.

The idea of the parallel planner is illustrated using the previous example in Fig. 2.6. For long-term planning, the robot plans a trajectory without considering the uncertainties in the prediction of the human movement. Then the long-term trajectory is used as a reference in the short-term planning. In the first time step in the short-term planner, the robot estimates the uncertainty in the prediction and checks whether it is safe to execute the long-term trajectory. As the trajectory does not intersect with the short-term uncertainty cone, it is executed. In the next step, since the trajectory is no longer safe, the short-term planner modifies the trajectory by detouring. Meanwhile, the long-term planner comes with another long-term plan and the previous plan is overwritten. The short-term planner then monitors the new reference trajectory. The robot follows the trajectory and finally approaches the goal. This approach addresses the uncertainty and is non-conservative. As a long-term planning module is included, the local optima problem in Fig. 2.5 can be avoided.

A block diagram of the parallel planning architecture is shown in Fig. 2.3. The computation time flow is shown in Fig. 2.7 together with the planning horizon and the execution horizon. Three long-term plans are shown, each with one distinct color. First, plan 1 (blue) is computed in the long-term planner. The upper part of the time axis shows the planning horizon. The middle layer is the execution horizon. Only part of the planned trajectory is executed. The bottom layer shows the computation time. The computation of the plan is done before the execution of the plan. Once computed, plan 1 is sent to the short-term planner for monitoring. The sampling rate in the short-term planner is much higher than that in the long-term planner, since the computation time in the short term planner is much smaller than that in the long term planner. The mechanism in the short term planner is similar to that in the long term planner. Note that the planning horizon in the short term planner is not necessarily one time step, though the execution horizon is one time step. While the short term planner is monitoring the trajectory, the long term planner is computing a new long

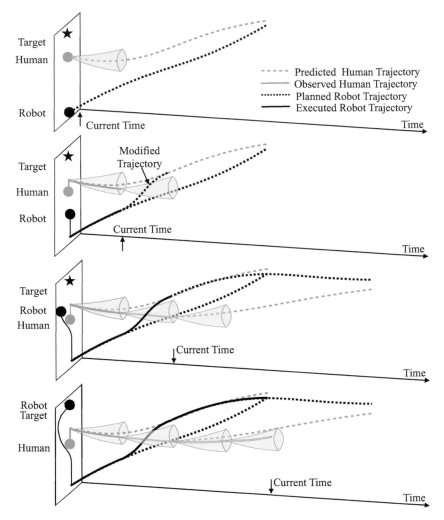

Figure 2.6: Illustration of the performance of the parallel planning.

term trajectory. Once the new trajectory is computed, it will be sent to the short-term planner for monitoring. The long-term planner then computes another trajectory and so on.

This approach can be regarded as a two-layer MPC approach. Coordination between the two layers is important. To avoid instability, a margin needs to be added to the safety constraint in the long-term planner so that the long-term plan will not be revoked by the short-term planner when the long-term prediction of the human motion is correct. Theoretical analysis of the stability of the two-layer MPC method is out of the scope of this book, and is left as a topic for future study.

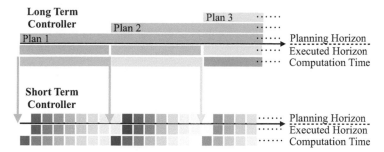

Figure 2.7: The time flow in the parallel planners.

Color version at the end of the book

As mentioned earlier, the long-term planning addresses the efficiency objective, which will be discussed in detail in Chapter 4. The short-term planning addresses the safety objective, which will be discussed in detail in Chapter 3.

Learning Process: The Cognition Skill

Agent i's model of the environment $\mathcal{M}_i$ needs to be updated online. Online update of the model is a cognition skill. The learning is divided into two parts: intention inference and model adaptation. For example, in a factory environment, intention inference tells whether a human worker is going to sit down, stand up, walk to a bench, or pick up a work piece. Indeed, intention inference identifies the context behind the observed motion. Mathematically, it can be understood as model selection. Given the intention (selected model), different agents may have different ways to realize the intention. For example, a cautious pedestrian may wait for all the vehicles to pass before he crosses the street, while another pedestrian may cross the street assuming that all the vehicles will yield. Hence model adaptation is needed to account for individual differences. Another reason for model adaptation is to account for time-varying behaviors. The system can be time-varying when all agents are adapting their behaviors to the environment. We can either build the structure of the models offline during the training phase or identify the models from sketch online. The architecture of the cognition system is shown in Fig. 2.8. The advantage of this architecture is that the prior knowledge in the design phase or in the training phase can be easily incorporated. The methods will be further explained in section 5.2.

2.2.2 MODEL-FREE DESIGN

We formulate the model-free design in two different frameworks: the imitation learning and the analogy learning. Both concern with the construction of a database, the design of the logic, and the design of the learning process for database augmentation.

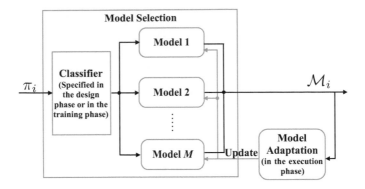

Figure 2.8: Cognition: the designed learning process in this book.

Knowledge: The Database Construction

Instead of applying models to explictly describe the system dynamics, model-free approaches utilize data to implicitly encode the system property. Constructing a high-quality and comprehensive database is a critical prerequisite for the following data mining process.

In general, two kinds of data, including scenario measurement $\pi \in \mathbb{R}^n$ and action $u \in \mathbb{R}^m$, will be recorded. Necessary filtering of the data is required to compensate the system disturbance or sensor noise. Dimension reduction can also be performed by projecting the raw data to a lower dimensional manifold.

Model-free Logic: Imitation or Analogy

Imitation learning is one of the classic model-free learning approaches. Figure 2.9 shows a general pipeline for imitation learning. During the training stage, the control policy $g : \mathbb{R}^n \to \mathbb{R}^m, u_{\text{train}} = g(\pi_{\text{train}})$ is regressed from the training data. The trained control policy is assumed to be universal and applicable, therefore it is applied to test states to achieve corresponding test actions $u_{\text{test}} = g(\pi_{\text{test}})$.

However, cognitive science finds that this pipeline is not necessarily the only way for human learning. Many studies showed that without understanding how the system progresses or the relation between states and actions (policy), humans can still make decisions through analogical reasoning [42, 164, 116]. Given a current scenario, human beings will try to find the correlation between the past and the current, and transfer previous problem-solving actions to apply to the current situation. This analogy approach does not require an understanding of the universal policy or causality behind the system. It also enables humans to learn efficiently from small data samples.

Based on this observation, an analogy-based learning approach is proposed in this book. Unlike those traditional model-free approaches which try to formulate control policies, this new approach focuses on finding correlations between scenar-

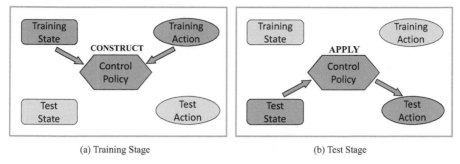

(a) Training Stage   (b) Test Stage

Figure 2.9: Pipeline of imitation learning. (a) During the training stage, construct a control policy which is a regression from measurements π to actions u. (b) During test stage, apply the control policy to new measurements π to achieve new actions u.

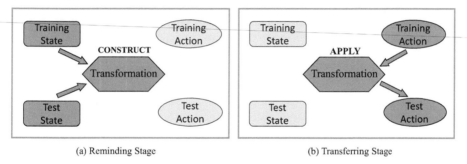

(a) Reminding Stage   (b) Transferring Stage

Figure 2.10: Pipeline of analogy learning. (a) During the training stage, construct a transformation between the correlated training measurement and test measurement. (b) During the transferring stage, apply the transformation to the training actions to achieve the test actions.

ios. As shown in Fig. 2.10, this approach has two general stages, the reminding stage and the transferring stage. During the reminding stage, the similarity between the past scenarios and the current scene will be calculated. The most similar pair will be selected and a mapping function between the two scenarios will be constructed: $g : \mathbb{R}^n \to \mathbb{R}^n, \pi_{\text{test}} = g(\pi_{\text{train}})$. During the transferring stage, the past actions will be transformed by the mapping function so as to generate new actions which are feasible for the current situation, i.e., $u_{\text{test}} = g(u_{\text{train}})$. More properties and advantages of analogy-based learning will be introduced in detail in Chapter 6.

Learning Process: Database Augmentation

Generalization is a key challenge for model-free approaches. Specifically, the policy is learned on the training data, while the actual test scenario might not be similar to any of the training scenarios. An life-long learning process is necessary to handle the unseen scenarios by constantly retraining and improving the policy.

To deal with this problem, Fig. 2.11 shows a classic learning process for imitation learning methods. Based on the constructed database (π_i, u_i), a temporary policy g is trained. Given a new test scenario π', the temporary policy is assumed to be generalizable and utilized to generate an action u'. The action is executed and an evaluation mechanism will evaluate the system to determine whether u' is appropriate or not. If not, the human operator will be asked to give a demonstration u' on the scenario π'. This new data pair (π', u') is augmented to the old database while a new iteration of training is performed to obtain an updated policy which can handle the scenario π'.

Similarly, Fig. 2.12 shows the learning procedures for analogy learning methods. A training database (π_i, u_i) which includes pairs of training states and corresponding actions, is given in advance as past experience. At test stage, the system arrives at a new state π'. Analogy learning will remind the agent the most similar scenario (π^*, u^*) in the database by calculating the similarity between the current scenario and past scenarios. A transformation function will be constructed to register π^* and π'. The same function is then utilized to transform the action u^* from the past scenario to get a new one u', which is feasible for the current scenario π'. If the similarity between π^* and π' is smaller than a threshold, which indicates that the most similar scenario in the past is still not similar enough with the current state, the transformation procedure will not be implemented since the base for analogy does not exist. Instead, human demonstration or traditional approaches will be utilized to provide the corresponding action u'. The demonstrated data sample (π', u') will then be augmented into the database. Since the database (experience) is growing, if the system runs into the similar scenario π' in the future, the analogy-based transformation can be executed without the need of additional demonstration.

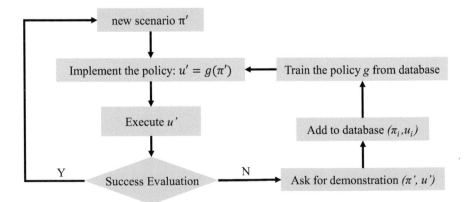

Figure 2.11: Procedures in imitation learning.

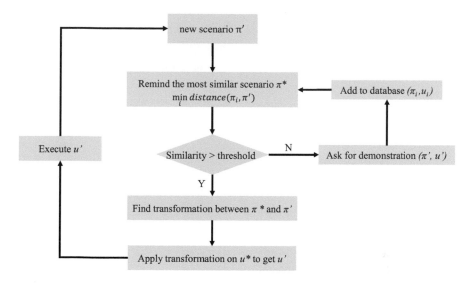

Figure 2.12: Procedures in analogy learning.

2.3   CONCLUSION

In this chapter, we modeled the human-robot interactions in a multi-agent framework. The considerations in robot behavior design were discussed and will be further elaborated in the following chapters. The application of the method will be discussed in Chapters 8 to 10.

Part II

Theory

3 Safety during Human-Robot Interactions

3.1 OVERVIEW

Human safety is one of the biggest concerns in HRI [204]. Two different approaches can be used to address the safety issues. One way is to increase the intrinsic safety level of the robot through hardware design or low level control, so that even if collision happens, the impact on the human is minimized [79]. The other way is to let the robot behave safely to prevent collision, which is called "interactive safety" as opposed to intrinsic safety [216]. In this chapter, interactive safety will be addressed in the context of local motion planning and control.

As discussed in section 1.4.1, new approaches to deal with safety in HRI are needed, which should be non-conservative and should ensure timely response of the robot. To properly design the robot behavior, the safety issues need to be understood in the human-robot systems, which belongs to multi-agent systems as discussed in Chapter 2. This multi-agent situation complicates the interactions as the individual optima may not align with the system optima [140], which is known as conflict of interests. An agent's interest is to be efficient, i.e., finish the task as soon as possible, while staying safe. For example, an automated vehicle's interest is to go to the destination in minimum time without colliding with surrounding vehicles. Safety is the mutual interests for all agents in the system, while the efficiency goals of different agents may conflict with one another. For example, for two vehicles during lane merging, they cannot pass the conflict zone at the same time. One of the vehicles must yield (i.e., sacrifice efficiency) to ensure safety.

Resolving conflicts among agents when there is no designated controlling agent is challenging. Chapter 7 will discuss several conflict resolution strategies. This chapter focuses on safety guarantee under conflicts. Take human's social behavior as an example, to ensure safety under conflicts of interests, a human's behavior is usually constrained. The constraint comes from at least two factors: the social norm and the uncertainties that he or she perceives for other people. The social norm guides human's behavior in a certain situation or environment as "mental representations of appropriate behavior". For example, the social norm will suggest an autonomous vehicle in a parking lot to keep a safe distance from pedestrians and detour if necessary. In practice, the perceived uncertainties also affect human's decision. Similarly, whether the autonomous vehicle needs to detour and how much the autonomous vehicle should detour highly depends on how certain the vehicle is about its prediction of the human's trajectory. Moreover, the uncertainties can be attenuated through active learning (refer to uncertainty reduction theory [25]). A common experience is: a newcomer tends to behave conservatively in a new environment due to large uncertainties. But through observing and learning from his peers, he will gradually behave freely due to the reduction of uncertainties.

We may encode similar behaviors through local planning for robots to be safe. Safety-oriented local planning fits into the model-based design in section 2.2.1. In the following discussion, we use subscript R to represent the robot and subscript H to represent all other agents. For simplicity, we refer to H as "the human". Suppose a reference control input u_R^o is received from the long-term planner of the robot, the short-term planner needs to ensure that the interaction constraint $x \in \mathcal{I}$ in (2.7d) will be satisfied after applying this input. The safe control problem can be formulated as the following optimization,

$$u_R^* = \arg\min_{u_R} \|u_R - u_R^o\|_Q^2 \tag{3.1a}$$

$$s.t. \ u_R \in \Omega, x \in \mathcal{I}, \dot{x}_R = f_R^x(x_R) + f_R^u(x_R)u_R, \tag{3.1b}$$

where $\|u_R - u_R^o\|_Q^2 := (u_R - u_R^o)^\mathsf{T} Q(u_R - u_R^o)$ penalizes the deviation from the reference input. The last equation (3.1b) is the robot dynamic function, which is assumed to be affine in the control input, where f_R^x and f_R^u are components in the dynamic function. The interaction constraint $x \in \mathcal{I}$ defines a safe set in the system state space. In this chapter, we use $\mathcal{X}_S$ to denote the safe set, which is equivalent to the interaction constraint $\mathcal{I}$. It is assumed that all agents agree on this constraint $\mathcal{X}_S$. The constraint can be understood as a social norm. The safe set and the robot dynamics impose non-linear and non-convex constraints which make the problem hard to solve. We will transform the non-convex state space constraint into a convex control space constraint using the idea of invariant set. The safe set algorithm (SSA) [130] is developed to enforce invariance in the safe set according to the predicted human behavior, and the safe exploration algorithm (SEA) further constrains the robot motion by uncertainties in the predictions [132]. By actively learning human behaviors, the robot will be more "confident" about its prediction of human motion, hence able to access a larger subset of the safe set when the uncertainty is smaller.

The remainder of the chapter is organized as follows: in section 3.2, the design methodology of the safety-oriented local motion planning will be discussed. The safe set algorithm (SSA) and the safe exploration algorithm (SEA) will be discussed in sections 3.3 and 3.4 respectively. A method to combine SSA and SEA will be discussed in section 3.5. Section 3.6 concludes the chapter.

3.2   SAFETY-ORIENTED BEHAVIOR DESIGN

Safety Principle

The safe set $\mathcal{X}_S$ poses a state space constraint R_S for the robot, where $R_S(x_H) = \{x_R : [x_R^\mathsf{T}, x_H^\mathsf{T}]^\mathsf{T} \in \mathcal{X}_S\}$. The constraint R_S is a projection of the safe set to the robot's state space given certain human state. If the human takes care of the

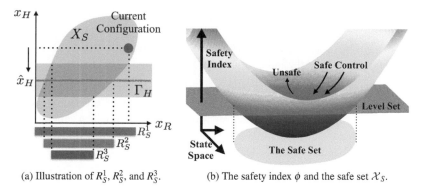

(a) Illustration of R_S^1, R_S^2, and R_S^3.   (b) The safety index ϕ and the safe set $\mathcal{X}_S$.

Figure 3.1: Illustration of the safe set, safety constraints, and the safety index.

safety, then the robot can go to any state as long as the human can find a way to stay safe

$$R_S^1 = \{x_R : x_R \in R_S(x_H) \text{ for some } x_H\}. \tag{3.2}$$

However, to make the system reliable, the safety problem should be taken care of by the robot. In case the robot knows the human's next move $\hat{x}_H$, the safety bound for the robot becomes

$$R_S^2 = \{x_R : x_R \in R_S(\hat{x}_H)\}. \tag{3.3}$$

Due to noises and uncertainties, the estimate $\hat{x}_H$ may not be accurate. The human state x_H may lie in a set Γ_H containing $\hat{x}_H$. Then the robot motion should be constrained in a smaller set

$$R_S^3 = \{x_R : x_R \in R_S(x_H), \forall x_H \in \Gamma_H\}. \tag{3.4}$$

Figure 3.1a illustrates the safe set $\mathcal{X}_S$ and the state space constraints R_S^1, R_S^2 and R_S^3. It is clear that $R_S^3 \subset R_S^2 \subset R_S^1$.

The Safety Principle: the robot control input $u_R(t)$ should be chosen such that $\mathcal{X}_S$ is time invariant, i.e., $x(t) \in \mathcal{X}_S$ for all t, or equivalently, $x_R(t) \in R_S^3(t)$ in (3.4) for some set $\Gamma_H(t)$ that accounts for almost all possible noises $v_1, ... v_N, w_1, ..., w_N$ and human behaviors $\mathcal{B}_i(\cdot)$ for $i \in H$ (those with negligible probabilities will be ignored).

Safety Index

The safety principle requires the designed control input to make the safe set invariant with respect to time. In addition to constraining the motion in the safe region R_S^3, the robot should also be able to cope with any unsafe human movement. Given the current configuration in Fig. 3.1a, if the human is anticipated to move downwards, the robot should go left in order for the combined trajectory to stay in the safe set. To cope with the safety issue dynamically, a safety index is introduced as shown in Fig. 3.1b. The safety index $\phi : \mathcal{X} \to \mathbb{R}$ is a function on the system state space such that

1. ϕ is differentiable with respect to t, i.e., $\dot{\phi} = (\partial\phi/\partial x)\dot{x}$ exists everywhere;
2. $\partial\dot{\phi}/\partial u_R \neq 0$;
3. The unsafe set $\mathcal{X}\backslash\mathcal{X}_S$ is not reachable given the control law $\dot{\phi} < 0$ when $\phi \geq 0$ and the initial condition $x(t_0) \in \mathcal{X}_S$.

The first condition ensures that ϕ is smooth. The second condition ensures that the robot input can always affect the safety index. The third condition provides a criterion to determine whether a control input is safe or not, e.g., all the control inputs that drive the state below the level set 0 are safe and unsafe otherwise.

Lemma: (Existence of the Safety Index) A function ϕ satisfying all three conditions exists for any set $\mathcal{X}_S$ that can be represented by a smooth function $\phi_0(x)$, i.e., $\mathcal{X}_S = \{x : \phi_0(x) \leq 0\}$.[1]

The safety index can be understood as a scalar energy function, which is designed such that the control objective (e.g., safety) is with low energy. The desired control law should make the system stay with low energy. There are other energy-function-based control methods. In those methods, the counterparts of the safety index are the potential function [168, 95], the barrier function [12], or the value function [68]. These energy functions may not necessarily satisfy the three conditions defined earlier. In general, the energy function (e.g., safety index) can be constructed manually or computed automatically using backward reachability analysis [154].

Safe Control

To ensure safety, the robot's control must be chosen from a set of safe control that makes the unsafe set unreachable, as specified in the third condition of the safety index. The set of safe control is denoted as $U_S(t) = \{u_R(t) : \dot{\phi} \leq -\eta_R \text{ when } \phi \geq 0\}$ where $\eta_R > 0$ is a safety margin. By the dynamic equation in (3.1b), the derivative of the safety index can be written as

$$\dot{\phi} = \frac{\partial\phi}{\partial x_R} f_R^u u_R + \frac{\partial\phi}{\partial x_R} f_R^x + \sum_{j\in H} \frac{\partial\phi}{\partial x_j}\dot{x}_j. \tag{3.5}$$

Then the set of safe control is

$$U_S(t) = \{u_R(t) : L(t)u_R(t) \leq S(t, \dot{x}_H)\}, \tag{3.6}$$

where

$$L(t) = \frac{\partial\phi}{\partial x_R} f_R^u, \tag{3.7}$$

$$S(t, \dot{x}_H) = \begin{cases} -\eta_R - \sum_{j\in H}\frac{\partial\phi}{\partial x_j}\dot{x}_j - \frac{\partial\phi}{\partial x_R}f_R^x & \phi \geq 0 \\ \infty & \phi < 0 \end{cases}. \tag{3.8}$$

---

[1]The Lemma is proved in [130]. ϕ can be constructed in the following procedure: first, check the order from ϕ_0 to u_R in the Lie derivative sense, denote it by n; then define ϕ as $\phi_0 + k_1\dot{\phi}_0 + ... + k_{n-1}\phi_0^{(n-1)}$. The coefficients $k_1,...,k_n$ are chosen such that the roots of $1 + k_1 s + ... + k_{n-1}s^{n-1} = 0$ all lie on the negative real line.

$L(t)$ is a vector at the "safe" direction, while $S(t, \dot{x}_H)$ is a scalar indicating how much control effort is needed to be safe. The scalar $S(t, \dot{x}_H)$ can be broken down into three parts: a margin $-\eta_R$, a term to compensate human motion $-\sum_{j \in H} \frac{\partial \phi}{\partial x_j} \dot{x}_j$ and a term to compensate the inertia of the robot itself $-\frac{\partial \phi}{\partial x_R} f_R^x$. In the following arguments when there is no ambiguity, $S(t, \dot{x}_H)$ denotes the value in the case $\phi \geq 0$ only. Under different assumptions of the human behavior, $S(t)$ varies. The sets of safe control correspond to R_S^1, R_S^2 and R_S^3 are

$$U_S^1(t) \quad = \quad \left\{ u_R(t) : L(t) u_R(t) \leq S(t, \dot{x}_H) \text{ for some } \dot{x}_H \right\}, \tag{3.9}$$

$$U_S^2(t) \quad = \quad \left\{ u_R(t) : L(t) u_R(t) \leq S(t, \hat{\dot{x}}_H) \right\}, \tag{3.10}$$

$$U_S^3(t) \quad = \quad \left\{ u_R(t) : L(t) u_R(t) \leq S(t, \dot{x}_H) \text{ for all } \dot{x}_H \in \dot{\Gamma}_H \right\}, \tag{3.11}$$

where $\hat{x}_H$ is the velocity vector that moves the current configuration of human x_H to the future configuration $\hat{x}_H$, and $\dot{\Gamma}_H$ is the set of velocity vectors that move x_H to the set Γ_H. Computationally, $\hat{\dot{x}}_H := \frac{\hat{x}_H - x_H}{t_s}$ where t_s is the sampling time. Obviously $U_S^3 \subset U_S^2 \subset U_S^1$. When the uncertainties in $\hat{\dot{x}}_H$ reduces, U_S^3 converges to U_S^2.

The difference between the two sets R_S and U_S is that R_S is static as it is on the state space, while U_S is dynamic as it concerns with the "movements". Due to the introduction of the safety index, the non-convex state space constraint $x_R \in R_S$ is transformed to a convex state space constraint $u_R \in U_S$. For example, consider the case that $x_R \in \mathbb{R}^2$ and $x_H \in \mathbb{R}^2$ represent positions of the robot and the human respectively. Let the safety constraint be such that the distance between the robot and the human is greater than a threshold $d_{min} > 0$, i.e., $\|x_R - x_H\|^2 \geq d_{min}$. Then the safe region R_S^3 defined by (3.4) is non-convex. But the set U_S^3 according to (3.11) is a half space when the system is on the boundary of the safe set, or the whole space when the system is inside the safe set.

To be compatible with the safety principle, the robot's control input should lie in the set U_S^3. There are two ways to choose the uncertainty bound $\dot{\Gamma}_H$. One way is to use a constant bound based on the mean prediction error of the human state, as will be introduced in Chapter 5. Another way is to dynamically adjust the uncertainty bound according to the level of uncertainties in real time. The safe set algorithm [130] adopts the first approach, while the safe exploration algorithm [132] adopts the second approach. These two algorithms will be discussed in the following two sections.

3.3   SAFE SET ALGORITHM

3.3.1   THE ALGORITHM

The safe set algorithm (SSA) always chooses the best control input in the set of safe control according to (3.11). The method to obtain the estimates $\hat{x}_H$ and $\hat{\dot{x}}_H$ will be discussed in section 5.2. A constant $\lambda_R^{SSA} > 0$ is introduced to bound the noises and uncertainties in the estimation. Hence the scalar S in (3.11) becomes

$$S^{SSA}(t, \hat{x}_H) = S(t, \hat{x}_H) - \lambda_R^{SSA} = -\eta_R - \lambda_R^{SSA} - \sum_{j \in H} \frac{\partial \phi}{\partial x_j} \hat{x}_j - \frac{\partial \phi}{\partial x_R} f_{Rx}^*. \qquad (3.12)$$

Then the set of safe control in SSA is

$$U_S^{SSA}(t) = \{u_R : L(t)u_R \le S^{SSA}(t, \hat{x}_H)\}. \qquad (3.13)$$

With (3.13), we can transform the non-convex optimization problem (3.1) into a convex optimization problem:

$$u_R^* = \min_{u_R \in \Omega \cap U_S^{SSA}} \|u_R - u_R^o\|_Q^2. \qquad (3.14)$$

It is assumed that Ω is convex. When the constraint Ω is not tight, we can solve the above optimization problem analytically. The safe control input $u_R^*(t)$ is

$$u_R^*(t) = u_R^o(t) - c\frac{Q^{-1}L(t)^{\mathsf{T}}}{L(t)Q^{-1}L(t)^{\mathsf{T}}}, \qquad (3.15)$$

where

$$c = \min_{u \in U_S^{SSA}(t)} |L(t)(u - u_R^o(t))|. \qquad (3.16)$$

3.3.2 EXAMPLE: PLANAR ROBOT ARM

In this example, we illustrate how to apply SSA on a planar robot arm, and demonstrate the effectiveness of the method through simulation.

The environment is shown in Fig. 3.2a where both the robot arm and the human need to approach their respective goals in minimum time without colliding with each other. The robot arm has two links with joint positions θ_1 and θ_2 and the joint velocities $\dot{\theta}_1$ and $\dot{\theta}_2$. Let $\theta = [\theta_1, \theta_2]^{\mathsf{T}}$. The control input is chosen to be the joint accelerations, i.e., $u_R = [\ddot{\theta}_1, \ddot{\theta}_2]^{\mathsf{T}}$. The end point position of the robot is denoted $p \in \mathbb{R}^2$. The relationship between the end point movement and the joint movement is

$$\dot{p} = J_p \dot{\theta}, \qquad (3.17)$$

where J_p is the Jacobian matrix of the robot arm at the end point p. Taking time derivative on both sides of (3.17), we obtain the relationship between the acceleration of the end point and the joint acceleration

$$\ddot{p} = J_p \ddot{\theta} + \dot{J}_p \dot{\theta}. \qquad (3.18)$$

Denote the robot's goal point in the work space as $g \in \mathbb{R}^2$. To make the robot go to the goal point, we can let the acceleration of the end point satisfy

$$\ddot{p} = K_p[p - g] + K_v \dot{p}, \qquad (3.19)$$

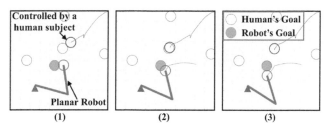

(a) The interaction between a robot arm and a human.

(b) The simulation result: scenario 1.

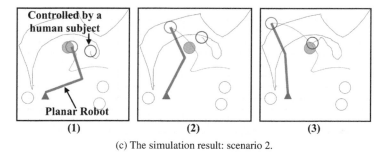

(c) The simulation result: scenario 2.

Figure 3.2: Application of SSA on a planar robot arm.

where $K_p \in \mathbb{R}^{2 \times 2}$ and $K_v \in \mathbb{R}^{2 \times 2}$ are the proportional and derivative control gains. By (3.18) and (3.19), the reference control input is designed to be

$$u_R^o = J_p^{-1} \left\{ K_p \left[p - g \right] + K_v \dot{p} - \dot{J}_p \dot{\theta} \right\}, \qquad (3.20)$$

The closest point to the human on the robot arm is denoted $m \in \mathbb{R}^2$. Define the robot state as $x_R = [m^\mathsf{T}, \dot{m}^\mathsf{T}]^\mathsf{T}$. The state space equation is

$$\dot{x}_R = A_R x_R + B_R J_m u_R + B_R H_m, \qquad (3.21)$$

where

$$A_R = \begin{bmatrix} 0 & \mathbf{I}_2 \\ 0 & 0 \end{bmatrix}, B_R = \begin{bmatrix} 0 \\ \mathbf{I}_2 \end{bmatrix}, \qquad (3.22)$$

and J_m is the Jacobian matrix at m with $H_m = \dot{J}_m \dot{\theta}$.

The human is simplified as a circle, whose state is taken as $x_H = [h^\mathsf{T}, \dot{h}^\mathsf{T}]^\mathsf{T}$ where $h \in \mathbb{R}^2$ is the human position and $\dot{h} \in \mathbb{R}^2$ is the velocity.

Define the safe set as $\mathcal{X}_S = \{x : d \geq d_{min}\}$ where $d = \|m - h\|_2$ measures the smallest ℓ_2 distance between the human and the robot arm and d_{min} is a positive constant. We need to design a safety index that satisfies the three conditions in section 3.2. Naively design the safety index as $\phi_0 = d_{min} - d$. However,

$$\frac{\partial \dot{\phi}_0}{\partial u_R} = -\frac{\partial d}{\partial x_R} B_R J_m = -\begin{bmatrix} \frac{\partial d}{\partial h} & \frac{\partial d}{\partial h} \end{bmatrix} \begin{bmatrix} 0 \\ \mathbf{I}_2 \end{bmatrix} J_m = 0. \tag{3.23}$$

The second condition in section 3.2 is violated. Indeed, we need to include the velocity term in the safety index. The safety index is then designed as

$$\phi = D - d^2 - k_\phi \dot{d}, \tag{3.24}$$

where $D = d_{min}^2 + \eta_R t_s + \lambda_R^{SSA} t_s$ and $k_\phi > 0$ are constants. It has been shown in [130] that the index satisfies all three conditions in section 3.2.

Given the safety index in (3.24), the parameters $L(t)$ and $S^{SSA}(t, \hat{x}_H)$ in the set of safe control (3.13) are derived below. Let the relative distance, velocity and acceleration vectors be $\mathbf{d} = m - h, \mathbf{v} = \dot{m} - \dot{h}$ and $\mathbf{a} = \ddot{m} - \ddot{h}$. Then $d = |\mathbf{d}|$ and

$$\begin{aligned}
\dot{\phi} &= -2d\dot{d} - k_\phi \ddot{d} = -2\mathbf{d}^\mathsf{T}\mathbf{v} - k_\phi \frac{\mathbf{d}^\mathsf{T}\mathbf{a} + \mathbf{v}^\mathsf{T}\mathbf{v} - \dot{d}^2}{d}, \\
&= -2\mathbf{d}^\mathsf{T}\mathbf{v} - k_\phi \frac{\mathbf{d}^\mathsf{T}(J_m u_R + H_m) - \mathbf{d}^\mathsf{T}\ddot{h} + \mathbf{v}^\mathsf{T}\mathbf{v}}{d} + k_\phi \frac{(\mathbf{d}^\mathsf{T}\mathbf{v})^2}{d^3}. \tag{3.25}
\end{aligned}$$

Hence,

$$L(t) = -k_\phi \frac{\mathbf{d}^\mathsf{T}}{d} J_m, \tag{3.26}$$

$$S^{SSA}(t, \hat{x}_H) = -\eta_R + 2\mathbf{d}^\mathsf{T}\mathbf{v} + k_\phi \frac{\mathbf{d}^\mathsf{T} H_m - \mathbf{d}^\mathsf{T}\hat{\ddot{m}} + \mathbf{v} \cdot \mathbf{v}}{d} - k_\phi \frac{(\mathbf{d}^\mathsf{T}\mathbf{v})^2}{d^3}, \tag{3.27}$$

where $\hat{x}_H = [\dot{m}^\mathsf{T}, \hat{\ddot{m}}^\mathsf{T}]^\mathsf{T}$. Then the computation of the safe control input follows from (3.15) and (3.16).

The simulation result is shown in Figs. 3.2b and 3.2c. In the simulation, several goals were assigned for the human. Before parameter adaptation, the robot first inferred the human's current goal [131]. Sampling time $t_s = 0.1s$. Figure 3.2b shows the human avoidance behavior of the robot. In (1), the human and the robot were both near their respective goals. However, since the human was heading towards the robot in high speed, the robot went backward in (2) and (3). Figure 3.2c shows the robot behavior under unexpected human behavior. In (1), the human suddenly changed his course. Although all of the human's goal points were in the lower part of the graph, the human started to go up. By observing that, the robot went away from the human in (2) and (3). The simulation results confirm the effectiveness of the algorithm.

3.3.3 EXAMPLE: VEHICLE

In this example, we illustrate the set U_S for vehicles in different scenarios, as shown in Fig. 3.3 to Fig. 3.4. For simplicity, the state of the vehicle is chosen as its position and velocity, while the control input is chosen as the acceleration. The dynamic equation of the vehicle is

$$\dot{x}_R = A_R x_R + B_R u_R. \tag{3.28}$$

The state x_H of the other vehicle includes its position and velocity. The safety index is chosen the same as in (3.24). Denote the relative distance, velocity, and acceleration vectors from the ego vehicle to the other vehicle as $\mathbf{d}$, $\mathbf{v}$, and $\mathbf{a}$. Then similar to (3.25) to (3.27), we have

$$
\begin{aligned}
\dot{\phi} &= -2d\dot{d} - k_\phi \ddot{d} = -2\mathbf{d}^\mathsf{T}\mathbf{v} - k_\phi \frac{\mathbf{d}^\mathsf{T}\mathbf{a} + \mathbf{v}^\mathsf{T}\mathbf{v} - \dot{d}^2}{d}, \\
&= -2\mathbf{d}^\mathsf{T}\mathbf{v} - k_\phi \frac{\mathbf{d}^\mathsf{T}u_R - \mathbf{d}^\mathsf{T}\ddot{h} + \mathbf{v}^\mathsf{T}\mathbf{v}}{d} + k_\phi \frac{(\mathbf{d}^\mathsf{T}\mathbf{v})^2}{d^3}, \tag{3.29}
\end{aligned}
$$

$$L(t) = -k_\phi \frac{\mathbf{d}^\mathsf{T}}{d}, \tag{3.30}$$

$$S^{SSA}(t, \hat{x}_H) = -\eta_R + 2\mathbf{d}^\mathsf{T}\mathbf{v} + k_\phi \frac{-\mathbf{d}^\mathsf{T}\hat{m} + \mathbf{v}\cdot\mathbf{v}}{d} - k_\phi \frac{(\mathbf{d}^\mathsf{T}\mathbf{v})^2}{d^3}. \tag{3.31}$$

According to (3.29) to (3.31), the set of safe control is affected by relative distance, relative velocity and predicted acceleration of the other vehicle. Figure 3.3 shows the safety constraint with respect to a front vehicle. The red dot represents zero acceleration and the circle represents the boundary of maximum acceleration in any direction (which may also in be other shapes, e.g., an ellipsoid). The shaded area represents U_S. When the front vehicle is relatively static with respect to the automated vehicle, U_S depends only on the relative distance. When the headway is far enough, all directions of acceleration are safe. When the headway is too short, only decelerations are safe. When the front vehicle has relative motions with respect to the

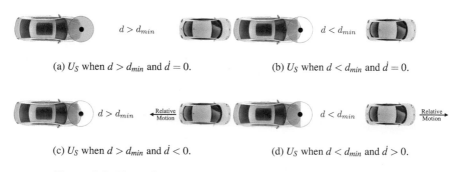

(a) U_S when $d > d_{min}$ and $\dot{d} = 0$.   (b) U_S when $d < d_{min}$ and $\dot{d} = 0$.

(c) U_S when $d > d_{min}$ and $\dot{d} < 0$.   (d) U_S when $d < d_{min}$ and $\dot{d} > 0$.

Figure 3.3: The safety constraint U_S with respect to a front vehicle.

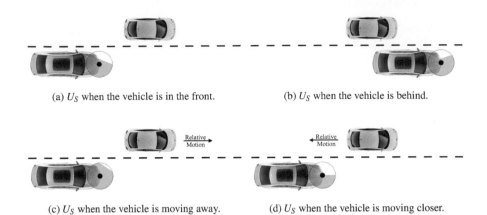

(a) U_S when the vehicle is in the front.   (b) U_S when the vehicle is behind.

(c) U_S when the vehicle is moving away.   (d) U_S when the vehicle is moving closer.

Figure 3.4: The safety constraint U_S with respect to a vehicle in the adjacent lane.

automated vehicle, U_S depends on the relative distance as well as relative movement. Figure 3.4 illustrates the safety constraint when there is a vehicle in the adjacent lane. In mixed traffic, the constraint U_S is the intersection of the constraint computed with respect to each surrounding vehicle, which will be further illustrated in section 8.3.3.

3.4   SAFE EXPLORATION ALGORITHM

In SSA, the bound for the uncertainties (i.e., λ_R^{SSA}) is a constant. However, the uncertainty of the human state is changing from time to time. A larger bound is needed if there is larger uncertainty. To capture this property, the safe exploration algorithm (SEA) is introduced, where the control input changes under different levels of uncertainties.

3.4.1   THE ALGORITHM

When there are uncertainties, the state estimate of x_j for $j \in H$ is no longer a point but a distribution, e.g., $\mathcal{N}(\hat{x}_j, X_j)$ where X_j is the covariance representing the level of uncertainties in the estimation. The collection of all those distributions is the belief space [56]. In this chapter, all distributions are assumed to be Gaussian. Since $x_j \sim \mathcal{N}(\hat{x}_j, X_j)$, the covariance can be written as

$$X_j = \mathbb{E}\left[(x_j - \hat{x}_j)(x_j - \hat{x}_j)^\top\right], \qquad (3.32)$$

which is called the mean squared estimation error (MSEE).

In discrete time system, for one-step look ahead prediction, we denote the estimate and MSEE of $x_j(k+1)$ at time step k as $x_j(k+1|k)$ and $X_j(k+1|k)$. The computation of the estimate and MESS will be discussed in section 5.2. Hence, at time step k, the best prediction for $x_j(k+1)$ has the following distribution

$$\mathcal{N}(\hat{x}_j(k+1|k), X_j(k+1|k)).$$ (3.33)

Since the distribution of $x_j(k+1)$ is unbounded, the inequality in (3.11) is ill-defined. Indeed, u_R needs to satisfy a probability constraint

$$\mathbb{P}\left(\{x_j(k+1) : L(k)u_R \leq S(k, x_H)\}\right) \geq 1 - \varepsilon, \forall j \in H,$$ (3.34)

where $\varepsilon > 0$ is a small number. The constraint (3.34) is also called a chance constraint. When ε is small enough, the inequality condition in (3.11) is almost satisfied. It is worth noting that enforcing the chance constraint (3.34) instead of the hard constraint (3.11) does not imply that the system may have ε probability to be unsafe. The assumption on the normal distribution of the prediction error is an approximation. This assumption is exploited to simplify the algorithm for human behavior estimation and prediction, to be discussed in section 5.2. The true distribution of prediction error may well be bounded. The chance constraint helps us to turn an unbounded probability distribution into a bounded set. The bounded set that satisfies the chance constraint (3.34) should be defined as $\Gamma_j(k)$ for all $j \in H$ such that

1. the probability density $\frac{d}{dx_j}\mathbb{P}(x_j)$ for $x_j \notin \Gamma_j(k)$ is small;
2. $\mathbb{P}(x_j \in \Gamma_j(k)) \geq 1 - \varepsilon$.

For a Gaussian distribution, the probability mass lying within the 3σ deviation is 0.997. Set $\varepsilon = 0.003$ and let $\Delta x_j = x_j - \hat{x}_j(k+1|k)$, then the set Γ_j can be defined as

$$\Gamma_j(k+1) = \left\{x_j : [x_j - \hat{x}_j(k+1|k)]^\mathsf{T} X_j(k+1|k)^{-1}[x_j - \hat{x}_j(k+1|k)] \leq 9\right\}.$$ (3.35)

Using the set Γ_j in (3.11), the constraint in U_R^3 is satisfied if the following inequality holds,

$$L(k)u_R \leq S^{SEA}(k) = \min_{x_j(k+1) \in \Gamma_j(k+1), \forall j \in H} \{S(k)\}.$$ (3.36)

The inequality in (3.36) is formulated in discrete time where $L(k)$ follows from (3.7) and $S(k)$ is similar to (3.8)

$$S(k) = -\eta_R - \sum_{j \in H} \frac{\partial \phi}{\partial x_j} \frac{x_j(k+1) - x_j(k)}{t_s} - \frac{\partial \phi}{\partial x_R} f_R^x.$$ (3.37)

Given (3.37), the minimization in (3.36) can be decoupled as a sequence of optimization problems, i.e., for all $j \in H$,

$$\min_{x_j(k+1) \in \Gamma_j(k)} \frac{\partial \phi}{\partial x_j} x_j(k+1).$$ (3.38)

By Lagrangian method[2], the optimal solution $x_j^*(k+1)$ for all $j \in H$ is

$$x_j^*(k+1) = \hat{x}_j(k+1|k) + \frac{3X_j(k+1|k)\left(\frac{\partial \phi}{\partial x_j}\right)^{\mathsf{T}}}{\left[\left(\frac{\partial \phi}{\partial x_j}\right)X_j(k+1|k)\left(\frac{\partial \phi}{\partial x_j}\right)^{\mathsf{T}}\right]^{\frac{1}{2}}}. \tag{3.39}$$

According to (3.39), $S^{SEA}(k)$ can be expressed as

$$S^{SEA}(k) = S(k,\hat{x}_H) - \lambda_R^{SEA}(k) = -\eta_R - \lambda_R^{SEA}(k) - \sum_{j \in H} \frac{\partial \phi}{\partial x_j}\hat{\dot{x}}_j - \frac{\partial \phi}{\partial x_R}f_{Rx}^*, \tag{3.40}$$

where

$$\lambda_R^{SEA}(k) = \frac{3}{t_s}\sum_{j \in H}\left[\left(\frac{\partial \phi}{\partial x_j}\right)X_j(k+1|k)\left(\frac{\partial \phi}{\partial x_j}\right)^{\mathsf{T}}\right]^{\frac{1}{2}} + \lambda_R^o, \tag{3.41}$$

and $\lambda_R^o \in \mathbb{R}^+$ is the bound for other uncertainties. Then the safe control signal can be computed following (3.15) and (3.16).

3.4.2   EXAMPLE: LOCAL PLANNING OF A VEHICLE

In this section, a comparative study between SSA and SEA is performed on a vehicle interacting with a human. The vehicle's state is denoted $x_R = [p^{\mathsf{T}}, v, \theta]^{\mathsf{T}}$ where $p \in \mathbb{R}^2$ is the position of the vehicle, $v \in \mathbb{R}$ the speed, and $\theta \in \mathbb{R}$ the direction. The control input of the vehicle is $u_R = [\dot{v}, \dot{\theta}]^{\mathsf{T}}$. The dynamic equation is

$$\dot{x}_R = \begin{bmatrix} v\cos\theta \\ v\sin\theta \\ 0 \\ 0 \end{bmatrix} + \begin{bmatrix} 0 \\ \mathbf{I}_2 \end{bmatrix} u_R. \tag{3.42}$$

The human's state is $x_H = [h^{\mathsf{T}}, \dot{h}^{\mathsf{T}}]^{\mathsf{T}}$, where $h \in \mathbb{R}^2$ is the position and $\dot{h} \in \mathbb{R}^2$ is the velocity of the human. Suppose that the goal point of the robot is $g \in \mathbb{R}^2$. Design a Lyapunov function $V = \|p - g\|^2 + v^2$, which is zero only when the robot stays at the goal point. To make the robot go to the goal point, the baseline control is designed

---

[2]The objective function is linear while the constraint function defines an ellipsoid. The optimal solution must lie on the boundary of the ellipsoid. Let $\gamma \in \mathbb{R}$ be a Lagrange multiplier. Define the new cost function as:

$$J_j^* = \frac{\partial \phi}{\partial x_j}x_j(k+1) + \gamma\left[9 - \Delta x_j^{\mathsf{T}}X_j(k+1|k)^{-1}\Delta x_j\right].$$

The optimal solution satisfies $\frac{\partial J_j^*}{\partial x_j(k+1)} = \frac{\partial J_j^*}{\partial \gamma} = 0$, i.e., $\left(\frac{\partial \phi}{\partial x_j}\right)^{\mathsf{T}} - 2\gamma X_j(k+1|k)^{-1}\Delta x_j = 0$ and $9 - \Delta x_j^{\mathsf{T}}X_j(k+1|k)^{-1}\Delta x_j = 0$. Then (3.39) follows.

to make $\dot{V} \leq 0$, i.e., always drive the Lyapunov function toward zero. One possible design of the baseline control is

$$\dot{v} = -(p-g)^\mathsf{T}\vec{e}_\theta - k_v v, \tag{3.43}$$

$$\dot{\theta} = k_\theta \left[\arctan(p-g) - \theta\right], \tag{3.44}$$

where $\vec{e}_\theta = [\cos\theta, \sin\theta]^\mathsf{T}$, $\arctan\vec{e}_\theta = \theta$, k_v and $k_\theta \in \mathbb{R}^+$ and are constants. It is easy to verify that $\dot{V} = -k_v v^2 \leq 0$ for the control law designed above.

The safety index $\phi = D - d^2 - k_\phi \dot{d}$ is chosen to be the same as in (3.24). In SSA, D is set to be $d_{min}^2 + \eta_R t_s + \lambda_R^{SSA} t_s$. In SEA, D is set to be $d_{min}^2 + \eta_R t_s + \lambda_R^{SEA}(k)t_s$. The relative distance, velocity and acceleration vectors are

$$\mathbf{d} = p - h,$$

$$\mathbf{v} = v\vec{e}_\theta - \dot{h},$$

$$\mathbf{a} = \dot{v}\vec{e}_\theta + v\vec{e}_{\theta+\pi/2}\dot{\theta} - \ddot{h}.$$

Similar to (3.25), the time derivative of the safety index is

$$\dot{\phi} = -2\mathbf{d}^\mathsf{T}\mathbf{v} - k_\phi \frac{\mathbf{d}^\mathsf{T}\mathbf{a} + \mathbf{v}^\mathsf{T}\mathbf{v} - \dot{d}^2}{d},$$

$$= -2\mathbf{d}^\mathsf{T}\mathbf{v} - k_\phi \frac{\left[\mathbf{d}^\mathsf{T}\vec{e}_\theta, \mathbf{d}^\mathsf{T}v\vec{e}_{\theta+\pi/2}\right]u_R - \mathbf{d}^\mathsf{T}\ddot{h} + \mathbf{v}^\mathsf{T}\mathbf{v}}{d} + k_\phi \frac{(\mathbf{d}^\mathsf{T}\mathbf{v})^2}{d^3}, \tag{3.45}$$

which implies

$$L(t) = k_\phi \left[\mathbf{d}^\mathsf{T}\vec{e}_\theta, \mathbf{d}^\mathsf{T}v\vec{e}_{\theta+\pi/2}\right], \tag{3.46}$$

$$S(t,\dot{x}_H) = -\eta_R + 2\mathbf{d}^\mathsf{T}\mathbf{v} + k_\phi \frac{\mathbf{v}^\mathsf{T}\mathbf{v} - \mathbf{d}^\mathsf{T}\ddot{h}}{d} - k_\phi \frac{(\mathbf{d}^\mathsf{T}\mathbf{v})^2}{d^3}. \tag{3.47}$$

Then S^{SSA} and S^{SEA} follow from (3.12) and (3.40) respectively, and the final control input follows from (3.15).

Figures 3.5a and 3.5b show the vehicle trajectories under SSA and SEA. The vehicle needed to approach $(0,5)$ from $(-5,-5)$ while the human went from $(0,-3)$ to $(-5,5)$. Five time steps are shown in the plots: $k = 3, 52, 102, 206, 302$ from the lightest to the darkest. The solid circles represent the human, which were controlled by a human subject through a multi-touch trackpad in real time. There was overshoot as the control was not perfect. The triangles represent the vehicle. The transparent circles in Fig. 3.5a represent the set $\Gamma_H(k)$ in (3.35) mapped into 2D, which is shrinking gradually due to the reduction of uncertainties as an effect of learning. In Fig. 3.5b, the transparent circles represent the equivalent uncertainty levels introduced by λ_R^{SSA}, thus the radius remains constant throughout the time.

Figure 3.6 shows the distance profiles and the vehicle velocity profiles under SSA and SEA. Due to large initial uncertainties, the vehicle only started to accelerate after $k = 50$ (when the relative distance was large) in SEA. However, in SSA, the vehicle tried to accelerate in the very beginning, then decelerated when the relative distance

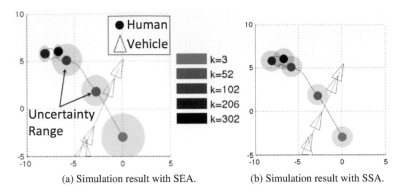

(a) Simulation result with SEA.   (b) Simulation result with SSA.

Figure 3.5: Application of the SEA algorithm and the SSA algorithms on an AGV.

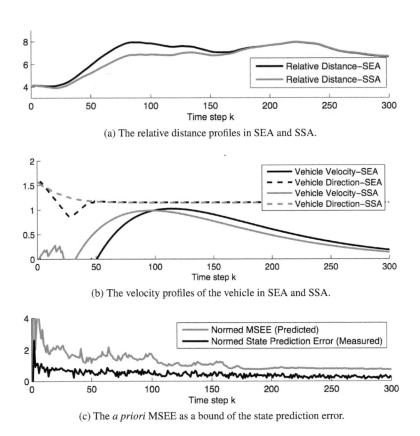

(a) The relative distance profiles in SEA and SSA.

(b) The velocity profiles of the vehicle in SEA and SSA.

(c) The *a priori* MSEE as a bound of the state prediction error.

Figure 3.6: Comparison between SEA algorithm and SSA algorithm.

to the human decreased. The velocity profile in SSA was serrated, while the one in SEA was much smoother. Meanwhile, in both algorithms, the relative distance was always greater than $d_{min} = 3$. However, before $k = 150$, the relative distance was kept larger in SEA than in SSA due to large uncertainty. Figure 3.6c shows that MSEE provides a perfect bound for the prediction error, while the prediction error reduces gradually, hence validates the learning algorithm to be discussed in section 5.2.

In conclusion, the behavior in SSA is: move and modify; while in SEA, it is: move only if confident. The behavior under SEA is better for a new comer, while the behavior under SSA is better if the robot is already very familiar with the environment, i.e., with low uncertainty levels.

3.5   AN INTEGRATED METHOD FOR SAFE MOTION CONTROL

In real world applications, the system topology is usually time varying, e.g., the robot will encounter different agents at different times in different locations [65]. Mathematically, that means some agents will be decoupled from the system block diagram in Fig. 2.2 and others will join from time to time. The robot is not faced with the "same" system throughout the time. This scenario is common for mobile robots and automated vehicles [134].

As the robot needs to deal with new agents, SEA is more appropriate than SSA. However, when the number of agents increase, the computational complexity in SEA increases dramatically. This section discusses a method to combine SSA and SEA in order to balance the performance and the computational complexity.

Due to limited sensing capability, the robot can only track humans that are within certain distance. Every agent within this range will be assigned a special identification number. Let $H(k)$ denote the collection of those identification numbers at time step k. A safety index ϕ_j is designed for each agent $j \in H(k)$. In this way, u_R needs to be constrained by

$$U_R^3 = \bigcap_{j \in H(k)} U_{R,j}^3 = \{u_R : L_j(k)u_R(k) \leq S_j(t, \dot{x}_j) \text{ for all } \dot{x}_j \in \dot{\Gamma}_j\}, \qquad (3.48)$$

where L_j and S_j are calculated with respect to ϕ_j. The uncertainty-bound $\dot{\Gamma}_j$ is chosen according to SEA only for new agents. Once the MSEE converges, the algorithm is switched to SSA for that agent. The idea is illustrated in algorithm 1, where Π is a set that records the identification numbers of the agents who are no longer considered as new agents.

3.6   CONCLUSION

This chapter discussed a general methodology to address safety in HRI. The safety issues were understood as conflicts in the multi-agent system. To solve the conflicts,

---

Algorithm 1 The Algorithm Combining SSA and SEA

Initialize $\Pi = \emptyset$, $k = 0$
while Controller is Active **do**
   $k = k + 1$
   Read current measurement and determine $H(k)$
   for $j \in H(k)$ **do**
      calculate the estimate $\hat{x}_j(k+1|k)$
      if $j \notin \Pi$ **then**
         calculate the MSEE $X_j(k+1|k)$
         if X_j converges **then**
            $\Pi = \Pi \cup \{j\}$
         end if
      end if
   end for
   for $j \in H(k)$ **do**
      if $j \in \Pi$ **then**
         $U_{R,j}^3 = U_{R,j}^{SSA} = \{u_R : L_j(k) \leq S_j^{SSA}(k)\}$ (Apply SSA to ϕ_j)
      else
         $U_{R,j}^3 = U_{R,j}^{SEA} = \{u_R : L_j(k) \leq S_j^{SEA}(k)\}$ (Apply SEA to ϕ_j)
      end if
   end for
   $U_R^3 = \bigcap_{j \in H(k)} U_{R,j}^3$
   Choose control u_R^* by optimizing over U_R^3
end while

---

the robot's behavior was constrained according to the "social norm" and the uncertainties it perceived for the other agents (including humans and other robots). Two algorithms were discussed under the framework: the safe set algorithm (SSA) and the safe exploration algorithm (SEA). In both algorithms, the robot calculated the optimal action to finish the task while staying safe with respect to the predicted human motion. The difference was that SEA actively tracked the uncertainty levels in the prediction and incorporated that information in the robot control, while SSA did not. As shown in the human-involved simulations, SEA was better when the uncertainty levels change from time to time, especially in the early stages of human-robot interactions. On the other hand, SSA was better when the predictions were already accurate, e.g., when the robot was "familiar" with the human, as SSA was more computationally efficient than SEA. Finally, a method to combine both algorithms was proposed to take advantage of both algorithms. Several case studies were presented and demonstrated the effectiveness of this method.

4 Efficiency in Real-Time Motion Planning

4.1 OVERVIEW

Global motion planning is one of the key challenges in robotics. Robots need to find motion trajectories to accomplish certain tasks in constrained environments in real time. The scenarios include but are not limited to navigation of unmanned aerial or ground vehicles in civil tasks such as search and rescue, surveillance and inspection; navigation of autonomous or driverless vehicles in future transportation systems; or motion planning for industrial robots in factory floor. In the context of human-robot interactions, the motion planning should be accomplished efficiently in real time in order for the robot to respond to the environmental changes. As discussed in sections 1.3.3 and 1.4.2, optimization-based motion planning methods can generate smoother trajectories compared to search-based or sampling-based methods, but have high computational complexity due to non-convexity of the problem. This chapter discusses how to improve motion efficiency using optimization-based methods, as well as how to efficiently solve the non-convex optimizations online.

Global motion planning is to find a trajectory that solves the constrained optimization problem (2.7) in a long time horizon. A trajectory has two attributions, spatial attribution and temporal attribution. To avoid dynamic obstacles such as humans, the robot can resort to either spatial maneuvers such as detours or temporal maneuvers such as slowing down or speeding up. Regarding these two kinds of maneuvers, there are two planning frameworks in literature, i.e., the integrated framework and the layered framework.

The *integrated* framework relies on spatiotemporal planning, which considers the spatial and temporal maneuvers simultaneously. On the other hand, the *layered* framework separates the considerations on the spatial and temporal maneuvers by planning a path first and then generating a speed profile along the path [71]. In the cases when there is not much freedom for the robot to choose alternative paths, nudging the speed profile is the only choice to respond to moving obstacles. Figure 4.1 illustrates the differences between the two frameworks under optimization-based methods. The horizontal plane represents a 2D state space. The vertical axis represents time. The shaded volume is the time-augmented obstacle. In the integrated framework, a trajectory needs to be computed directly in the spatiotemporal space starting from a simple reference trajectory connecting the start point and the target point, as shown in Fig. 4.1a. In the layered framework, the trajectory is planned by assigning temporal information to the points on a given path as shown in Fig. 4.1b.

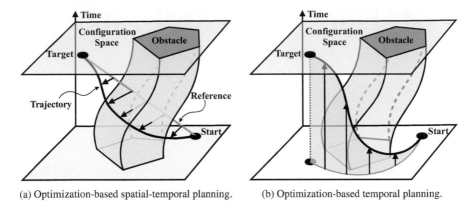

(a) Optimization-based spatial-temporal planning. (b) Optimization-based temporal planning.

Figure 4.1: Optimization-based motion planning in integrated and layered frameworks.

In the remainder of this chapter, the global motion planning problem will be reformulated first in section 4.2, followed by a discussion of optimization-based trajectory planning in an integrated framework in section 4.3, optimization-based speed profile planning in section 4.4, and integrated optimization-based layered planning in section 4.5.

4.2 PROBLEM FORMULATION

This section reformulates the problem (2.7) in the context of trajectory planning.

Trajectory Representation

Denote the configuration of the robot as $p \in \mathcal{C} \subset \mathbb{R}^n$ where $\mathcal{C}$ is an n dimensional configuration space. A configuration is a complete specification of the position of every point in the robotic system. The configuration of a robot may not be identical to the state x_R of the robot since configuration does not include velocity information. For a mobile robot, p is the position of the robot in the plane; for an aerial robot, p is the position of the robot in space; for a robot arm, p is the joint position of the robot.

Let $\mathbf{p} : [0, s^*] \rightarrow \mathcal{C}$ denote a path which is parameterized by length s. The total length of the path is s^*. $\mathbf{p}(s) \in \mathbb{R}^n$ is a point on the path $\mathbf{p}$ which has distance s to the starting point $\mathbf{p}(0)$. The time derivatives $\dot{\mathbf{p}} := \partial \mathbf{p} / \partial s$ and $\ddot{\mathbf{p}} := \partial^2 \mathbf{p} / \partial s^2$ denote the tangent vector and the normal vector of the path respectively. Since the path $\mathbf{p}$ is parameterized by length, its tangent vector has unit length, i.e., $\|\dot{\mathbf{p}}\| = 1$, and the length of its normal vector represents the curvature of the path. Hence we define the curvature $\kappa : [0, s^*] \rightarrow \mathbb{R}$ along the path $\mathbf{p}$ as $\kappa(s) := \|\ddot{\mathbf{p}}(s)\|$.

Let $\mathbf{s} : [0, t^*] \to [0, s^*]$ be a mapping from time to stations on the path. $\mathbf{s}$ encodes the speed profile of the path $\mathbf{p}$, i.e., $\dot{\mathbf{s}}(t)$ is the speed of the object at time t. This mapping $\mathbf{s}$ is non decreasing, but may not be bijective as the speed can be zero.

With the speed profile, a trajectory $\mathbf{T} : [0, t^*] \to \mathbb{R}^n$ is defined to be $\mathbf{T} := \mathbf{p} \circ \mathbf{s}$ where $\circ$ denotes function composition. The velocity along the path is

$$\dot{\mathbf{T}}(t) = \dot{\mathbf{p}}(\mathbf{s}(t)) \cdot \dot{\mathbf{s}}(t), \tag{4.1}$$

which equals to the speed $\dot{\mathbf{s}}$ times the tangent vector $\dot{\mathbf{p}}$ at point $\mathbf{s}(t)$. The acceleration along the path is

$$\ddot{\mathbf{T}}(t) = \ddot{\mathbf{p}}(\mathbf{s}(t)) \cdot \dot{\mathbf{s}}(t)^2 + \dot{\mathbf{p}}(\mathbf{s}(t)) \cdot \ddot{\mathbf{s}}(t), \tag{4.2}$$

which has two components, i.e., the longitudinal acceleration $\ddot{\mathbf{s}}$ and the lateral acceleration $\kappa(\mathbf{s})\dot{\mathbf{s}}^2$. For simplicity, we sometimes omit the parameter t or s if there is no ambiguity.

Let $\mathbf{p}_{a \to b}$ be a path connecting points a and b in $\mathbb{R}^n$, i.e., $\mathbf{p}(0) = a$ and $\mathbf{p}(s^*) = b$. Similarly, let $\mathbf{s}_{c \to d}$ be a speed profile such that $\dot{\mathbf{s}}(0) = c$ and $\dot{\mathbf{s}}(t^*) = d$. Let $\mathbf{T}_{[a,c] \to [b,d]}$ be a trajectory $\mathbf{T}$ composed of $\mathbf{p}_{a \to b}$ and $\mathbf{s}_{c \to d}$. To make a trajectory realizable, it needs to satisfy several constraints, i.e., safety constraint, dynamic constraint, boundary constraint.

Safety Constraint

The area in the Cartesian space that is occupied by the robot with configuration p is denoted as $\mathcal{V}(p) \in \mathbb{R}^k$ where $k = 2$ or 3 is the dimension of the Cartesian space. The area occupied by the obstacles or other agents in the environment at time t is denoted as $\mathcal{E}_t \in \mathbb{R}^k$. The area $\mathcal{E}_t$ for future time t can be predicted using methods to be discussed in section 5.2. Let $d_E : \mathbb{R}^k \times \mathbb{R}^k \to \mathbb{R}$ be the Euclidean distance function in the Cartesian space. The Euclidean distance from the robot to the obstacles is computed as

$$d(p, \mathcal{E}_t) := \min_{y \in \mathcal{V}(p), z \in \mathcal{E}_t} d_E(y, z). \tag{4.3}$$

Suppose the minimum distance requirement is d_{min}. Then the obstacle in the configuration space at time t is

$$\mathcal{O}_t := \{ p : d(p, \mathcal{E}_t) \leq d_{min} \}. \tag{4.4}$$

Note that $\mathcal{O}_t$ can be regarded as the projection of the Cartesian space obstacle $\mathcal{E}_t$ to the configuration space. An illustration of $\mathcal{E}_t$ and $\mathcal{O}_t$ for a planar robot arm is shown in Fig. 4.2. The safety constraint on the trajectory is

$$\mathbf{T}(t) \notin \mathcal{O}_t, \forall t. \tag{4.5}$$

Dynamic Constraint

The dynamic constraint comes from kinematic limits and limitations on control effort. Kinematic limits include speed limits and acceleration limits. For example, the curvature of the path should be bounded for non-holonomic vehicles. Denote the constraint as

$$F(\mathbf{T}, \dot{\mathbf{T}}, \ddot{\mathbf{T}}, \cdots) \in \Omega \tag{4.6}$$

where F is a function that computes the dynamic features, e.g., maximum speed, maximum curvature, maximum control input, etc. The set Ω encodes the bounds for those features.

Boundary Constraint

The initial condition on a trajectory is always fixed as it is determined by the current position and velocity of the robot. The terminal condition can vary depending on the applications. We can either require the robot to satisfy some condition on position and velocity at certain time (fixed time horizon planning), or require the robot to enter a goal state eventually (fixed target planning). One example of fixed time horizon problems is vehicle lane following, in which case the vehicle is only required to track the lane center at a desired speed, instead of going to a target position. One example of fixed target problems is robot grasping or vehicle parking which require the robot to go to certain destination.

The Planning Problem

With previous discussion, the global motion planning problem can be re-formulated as

$$\min_{\mathbf{T}_{\mathcal{A} \to \mathcal{Z}}} \quad J(\mathbf{T}_{\mathcal{A} \to \mathcal{Z}}) \tag{4.7a}$$

$$F(\mathbf{T}, \dot{\mathbf{T}}, \ddot{\mathbf{T}}, \cdots) \in \Omega \tag{4.7b}$$

$$\mathbf{T}(t) \notin \mathcal{O}_t, \forall t \tag{4.7c}$$

where (4.7a) is the cost function, (4.7b) is the dynamic constraint and (4.7c) is the safety constraint. $\mathcal{A}$ and $\mathcal{Z}$ are the boundary conditions. $\mathcal{Z}$ can encode either fixed target constraint or fixed time horizon constraint.

4.3   OPTIMIZATION-BASED TRAJECTORY PLANNING

Optimization methods are widely adopted in robot trajectory planning due to their flexibility to address multiple objectives [177, 189]. The major challenges lie in real-time computation. The optimization problem for trajectory smoothing in a clustered environment is usually highly non-convex, which is hard to solve in real time using conventional non-convex optimization solvers such as sequential quadratic programming (SQP) [29]. A convex feasible set algorithm (CFS)[127] is proposed to convexify the problems. This section first discusses the optimization-based trajectory planning problem in the framework of (4.7), followed by a quadratic approximation of the non-convex problem using CFS. Then its performance will be illustrated through several examples. For simplicity, this section only considers fixed time horizon planning.

4.3.1 PROBLEM FORMULATION

Discretize the trajectory $\mathbf{T}$ into h points and define $x_q = \mathbf{T}(qt_s)$ where t_s is the sampling time. The discrete trajectory is now denoted as $\mathbf{x} = [x_0^T, x_1^T, \cdots, x_h^T]^T$. A reference trajectory is denoted as $\mathbf{T}^r$ and also discretized as $\mathbf{x}^r$ which consists of a sequence of reference states $x_q^r = \mathbf{T}^r(qt_s)$ for different time steps q. The reference trajectory can be obtained from search-based or sampling based methods or can be simply constructed by connecting the two boundary sets. Define the finite difference operators $V \in \mathbb{R}^{nh \times n(h+1)}$ and $A \in \mathbb{R}^{n(h-1) \times n(h+1)}$ as

$$V = \frac{1}{t_s}\begin{bmatrix} \mathbf{I}_n & -\mathbf{I}_n & 0 & \cdots & 0 \\ 0 & \mathbf{I}_n & -\mathbf{I}_n & \cdots & 0 \\ \vdots & \vdots & \vdots & \ddots & \vdots \\ 0 & 0 & \cdots & \mathbf{I}_n & -\mathbf{I}_n \end{bmatrix}, A = \frac{1}{t_s^2}\begin{bmatrix} \mathbf{I}_n & -2\mathbf{I}_n & \mathbf{I}_n & 0 & \cdots & 0 \\ 0 & \mathbf{I}_n & -2\mathbf{I}_n & \mathbf{I}_n & \cdots & 0 \\ \vdots & \vdots & \vdots & \ddots & \ddots & \vdots \\ 0 & 0 & \cdots & \mathbf{I}_n & -2\mathbf{I}_n & \mathbf{I}_n \end{bmatrix}$$

Note that $V\mathbf{x}$ is the velocity vector and $A\mathbf{x}$ is the acceleration vector of the trajectory $\mathbf{x}$. Rewriting (4.7) in the discrete time, we have

$$\min_{\mathbf{x}} J(\mathbf{x} \mid \mathbf{x}^r) = w_1 \|\mathbf{x} - \mathbf{x}^r\|_Q^2 + w_2 \|\mathbf{x}\|_S^2 \tag{4.8a}$$

$$s.t.\ x_0 \in \mathcal{A}, x_h \in \mathcal{Z} \tag{4.8b}$$

$$\underline{\mathbf{v}} \le V\mathbf{x} \le \bar{\mathbf{v}}, \underline{\mathbf{a}} \le A\mathbf{x} \le \bar{\mathbf{a}} \tag{4.8c}$$

$$x_q \notin \mathcal{O}_q, \forall q = 1, \cdots, h \tag{4.8d}$$

The cost function is designed to be quadratic where $w_1, w_2 \in \mathbb{R}^+$. $\|\mathbf{x} - \mathbf{x}^r\|_Q^2 := (\mathbf{x} - \mathbf{x}^r)^T Q(\mathbf{x} - \mathbf{x}^r)$ penalizes the distance from the target trajectory to the reference trajectory. $\|\mathbf{x}\|_S^2 := \mathbf{x}^T S\mathbf{x}$ penalizes the properties of the target trajectory itself, e.g., length of the trajectory and magnitude of acceleration. The positive definite matrices $Q, S \in \mathbb{R}^{n(h+1) \times n(h+1)}$ can be constructed from the following components:

1. matrix for position $Q_1 = \mathbf{I}_{n(h+1)}$
2. matrix for velocity $Q_2 = V^T V$
3. matrix for acceleration $Q_3 = A^T A$.

Then $Q = \sum_{i=1}^{3} c_i^q Q_i$ and $S = \sum_{i=1}^{3} c_i^s Q_i$ where c_i^q and c_i^s are positive constants. Constraint (4.8b) is the boundary condition. Constraint (4.8c) corresponds to the dynamic constraint, which requires that velocities and accelerations along the trajectory are bounded by $\underline{\mathbf{v}}, \bar{\mathbf{v}} \in \mathbb{R}^{nh}$ and $\underline{\mathbf{a}}, \bar{\mathbf{a}} \in \mathbb{R}^{n(h-1)}$. Constraint (4.8d) corresponds to the safety constraint. The problem is non-convex due to (4.8d).

4.3.2   QUADRATIC APPROXIMATION

The non-convex optimization problem (4.8) is hard to solve in general using conventional non-convex optimization solvers. We will solve it using CFS, which transforms problem (4.8) into a quadratic program. In the formulation (4.8), the non-convexity only comes from the safety constraint (4.8d). We discuss methods to convexify it below. In other problems, it is also possible for the dynamic constraint (4.8c) to be non-convex. For example, the dynamic constraint can be non-convex when the curvature of the trajectory is constrained, since curvature depends non-linearly on $\mathbf{x}$. A method to convexify dynamic constraints is discussed in [136].

Convex Feasible Set

According to (4.8d), at each time step q, the infeasible set is $\mathcal{O}_q$. Let $d_E^* : \mathbb{R}^n \times \mathbb{R}^n \to \mathbb{R}$ be the Euclidean distance function in the configuration space. Then constraint (4.8d) is equivalent to $d^*(x_q, \mathcal{O}_q) \geq 0$, where $d^*(x_q, \mathcal{O}_q)$ is the signed distance function to $\mathcal{O}_q$ such that

$$d^*(x_q, \mathcal{O}_q) = \begin{cases} \min_{z \in \partial \mathcal{O}_q} d_E^*(x_q, z) & x_q \notin \mathcal{O}_q \\ -\min_{z \in \partial \mathcal{O}_q} d_E^*(x_q, z) & x_q \in \mathcal{O}_q \end{cases} \tag{4.9}$$

where $\partial \mathcal{O}_q$ denotes the boundary of the obstacle $\mathcal{O}_q$. Figure 4.2 illustrates the obstacle in the Cartesian space and the corresponding $\mathcal{O}_q$ (shaded area) in the configuration space for a planar robot arm.

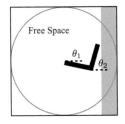
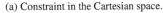
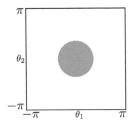

(a) Constraint in the Cartesian space.   (b) Constraint in the Configuration space.

Figure 4.2: Constraints for a planar robot arm in Cartesian and Configuration spaces.

The function $d^*(\cdot, \mathcal{O}_q)$ is convex, if $\mathcal{O}_q$ is convex. Assume $\mathcal{O}_q$ is convex. Then we can safely replace the nonlinear constraint $d^*(x_q, \mathcal{O}_q) \geq 0$ with a linear constraint $d^*(x_q^r, \mathcal{O}_q) + \nabla d^*(x_q^r, \mathcal{O}_q)(x_q - x_q^r) \geq 0$, since

$$d^*(x_q, \mathcal{O}_q) \geq d^*(x_q^r, \mathcal{O}_q) + \nabla d^*(x_q^r, \mathcal{O}_q)(x_q - x_q^r) \tag{4.10}$$

The linear constraint is a strict subset of the original constraint. Hence, the new constraint is safe but more conservative. If the obstacle $\mathcal{O}_q$ is non-convex, we can break it into several simple convex objects $\mathcal{O}_q^i$ such as circles or spheres, polygons or polytopes. The $\mathcal{O}_q^i$'s need not be disjointed. Then $d^*(\cdot, \mathcal{O}_q^i)$ is the convex cone of the convex set $\mathcal{O}_q^i$ as shown in Fig. 4.3a. Replacing (4.8d) with its first order approximation, the new optimization problem becomes

$$\min_{\mathbf{x}} \; J(\mathbf{x}; \mathbf{x}^r) = w_1 \|\mathbf{x} - \mathbf{x}^r\|_Q^2 + w_2 \|\mathbf{x}\|_S^2 \tag{4.11a}$$

$$s.t. \; x_0 = \mathcal{A}, x_h = \mathcal{Z} \tag{4.11b}$$

$$v_{min} \leq V\mathbf{x} \leq v_{max}, a_{min} \leq A\mathbf{x} \leq a_{max} \tag{4.11c}$$

$$d^*(x_q^r, \mathcal{O}_q^i) + \nabla d^*(x_q^r, \mathcal{O}_q^i)(x_q - x_q^r) \geq 0, \forall q, i \tag{4.11d}$$

The constraint (4.11d) is a convex feasible set of the original non-convex constraint since $d^*(\cdot, \mathcal{O}_q^i)$ are convex for all i. The non-convex problem (4.8) reduces to a quadratic problem (4.11) which can be solved efficiently online. It has been shown in [127] that the constraint (4.11d) is always feasible if $\mathcal{O}_q^i$'s are disjoint with each other for all i. The constraint (4.11d) may be infeasible for x_q if $\mathcal{O}_q^i$'s overlap with one another. The feasibility problem will be left as future work.

For iterative implementation, we can set $\mathbf{x}^{(0)}$ to be $\mathbf{x}^r$ and substitute x_q^r in (4.11d) with $x_q^{(k)}$ at each iteration k. It has been shown in [127] that the sequence $\mathbf{x}^{(k)}$ generated by the iterations will converge to a local optimum of the original problem (4.8). Nonetheless, it will be shown that the trajectory after one iteration is good enough in the sense of feasibility and optimality. Thus, we can safely work with the non-iterative version if strict optimality is not required and the computation time is limited.

Visualization of the Convex Feasible Set

Figure 4.3a illustrates a convex feasible set in (4.11d)) for one q and i when $X \subset \mathbb{R}^2$. The left hand side of (4.11d) represents the tangent plane of the distance function d^* at the reference point x_q^r. Due to the cone structure of d^*, the tangent plane touches the boundary of the obstacle. The convex feasible set is the projection of the positive portion of the tangent plane onto $\mathbb{R}^2$, which is a half space. The half space is maximal in the sense that the distance from the reference point to the boundary of the half space is maximized, which is equal to the distance from the reference point to the obstacle. With this observation, we can construct the convex feasible set using a purely geometric method without differentiation. For any reference point x^r and any convex obstacle $\mathcal{O}$, denote the closest point on $\mathcal{O}$ to x^r as b^*. The convex feasible set for x^r with respect to $\mathcal{O}$ is just the half space that goes through b^* and whose normal direction is along $b^* - x^r$ as shown in Fig. 4.3b.

At each time step q, the convex feasible set with respect to all obstacles is a polygon, since it is the intersection of all feasible half spaces. Thus the convex feasible set for the whole planning horizon is a "tube" around the reference trajectory whose cross sections are those polygons. Similar ideas of using a "tube" constraint to simplify the trajectory smoothing problem can be found in [234]. The advantages of CFS

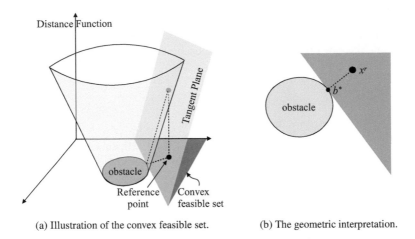

(a) Illustration of the convex feasible set.   (b) The geometric interpretation.

Figure 4.3: The convex feasible set and its geometric interpretation.

over the other methods are that (1) the "tube" is maximized and (2) the feasibility and convergence of the algorithm is guaranteed theoretically.

4.3.3 EXAMPLES: TRAJECTORY PLANNING FOR VARIOUS SYSTEMS

In this section, trajectory planning problems for different robots are considered, including a mobile robot and a planar robot arm. The optimization-based planning problem is solved using both CFS and SQP. The algorithms are run in MATLAB® (using MATLAB® script) on a MacBook of 2.3 GHz using Intel Core i7. SQP solves the problem (4.8) directly using MATLAB® `fmincon` function. CFS transforms the problem to (4.11) and solves it using MATLAB® `quadprog` function iteratively until convergence. The termination conditions for the two algorithms are set to be the same. For CFS, we record both the processing time for transforming the problem from (4.8) to (4.11) as well as the computation time for the resulting quadratic problem.

Mobile Vehicle

Consider navigation of mobile vehicles in indoor environments or autonomous driving in parking lots as shown in Fig. 4.4. In this case, $X = \mathbb{R}^2$ and $\mathcal{O}_q$ is a static maze for all q. We identify the configuration space with the Cartesian space and assume that the nonlinear constraint on vehicle dynamics is considered in the reference trajectory and will not be violated if the reference trajectory is slightly modified. The maze is first partitioned into five polygons as shown in Fig. 4.4a. The reference trajectory (shown as the solid line in Fig. 4.4a) is computed using Lattice A* search [137].

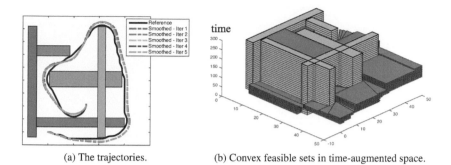

(a) The trajectories.   (b) Convex feasible sets in time-augmented space.

Figure 4.4: Trajectory smoothing for a mobile vehicle.

Since the possible turning directions are discretized, there are undesirable oscillations in the reference trajectory. In the optimization problem, a large penalty on acceleration is applied in order to get rid of the oscillations. The convex feasible set in (4.11d) is illustrated in a time-augmented state space in Fig. 4.4b where the z-axis represents time. At each time step, the convex feasible set is just a polygon around the reference point. All those polygons form a "tube" around the reference trajectory. The horizon of the problem is $h = 116$. The dimension of the optimization problem is 234. The safety margin is required to be $d_{min} = 3$.

The CFS algorithm converges after 5 iterations, with total computation time 0.9935 s. The smoothed trajectories for each iteration are shown in Fig. 4.4a. The average time for transforming the problem from (4.8) to (4.11) during each iteration is 0.1632 s. The average time for solving the optimization problem (4.8) is 0.0355 s. The cost profile from $J(\mathbf{x}^r)$ to $J(\mathbf{x}^{(5)})$ is

$$423.7 \rightarrow 584.3 \rightarrow 455.0 \rightarrow 407.6 \rightarrow 403.7 \rightarrow 403.6.$$

It is worth noting that although the cost changes quite a lot from the first iteration, the resulting paths are similar to each other as shown in Fig. 4.4a. The descent in cost is due to the adjustment of the velocity and acceleration profiles, i.e., redistributing sample points on the path.

SQP does not converge or even find a feasible solution within the max function evaluation limit 5000 with computation time 387.2 s. SQP undergoes 20 iterations. When terminated, the cost drops from 423.7 to 287.2, but the feasibility of the trajectory, i.e., $-\min_{q,i} d^*(x_q, \mathcal{O}_q^i)$, only goes from 3.1 to 2.3. The trajectory is feasible if and only if the feasibility value is less than or equal to zero.

Robot Arm

In this case, a three-link planar robot arm is considered, as shown in Fig. 4.5 with $X = [0, 2\pi)^3$. The infeasible area in the Cartesian space is wrapped by a capsule,

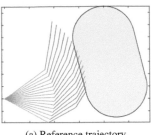

(a) Reference trajectory.

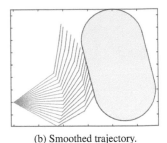

(b) Smoothed trajectory.

Figure 4.5: Trajectory smoothing for a robot arm.

which characterizes $\mathcal{O}_q$ for all q. The reference trajectory is generated by linear interpolation between boundary conditions $\mathcal{A}$ and $\mathcal{Z}$, which violated the constraint. The horizon is $h = 15$.

CFS converges after 2 iterations, with total computation time 2.4512 s. The smoothed trajectory is shown in Fig. 4.5b. The average time for transforming the problem from (4.8) to (4.11) during each iteration is 1.2213 s while the average time for solving the optimization problem is 0.0043 s. The pre-processing time is much longer than that for the mobile vehicle due to the nonlinearity of the robot arm. Moreover, the solution in the first iteration is already good enough and the second iteration is just for verification that a local optimum has been found.

In comparison, SQP converges after 41 iterations, with total computation time 5.34 s. Both CFS and SQP converge to the local optima at $J = 616.7$.

Scalability of the Algorithms

To further illustrate the advantage and scalability of CFS, we compared CFS with SQP and interior point method (ITP) in C++ on a simplified 2D mobile robot motion planning problem. As shown in Fig. 4.6, a mobile robot started at $(0,0)$ and tried to follow the straight purple dash reference trajectory. However, the green obstacle blocked the reference trajectory, which required a motion planner to modify the path. In this scenario, CFS, ITP and SQP were implemented in Knitro [38] in C++. In order to further analyze the performance among various methods, the same problem was solved with different planning horizons, where the number of steps went from 10 to 100. Table 4.1 summarizes the total computation time and the computation time per iteration respectively. The computation time of SQP was untraceable when the planning horizon increased to 80; hence, the corresponding results were not listed.

In terms of total computation time, CFS always outperforms ITP and SQP, since it requires less time per iteration and fewer iterations to converge. This is due to the fact that CFS does not require additional line search after solving (4.11) as is needed in ITP and SQP, hence saving time during each iteration. CFS requires fewer iterations to converge since it could take unconstrained step lengths $||\mathbf{x}^{(k+1)} - \mathbf{x}^{(k)}||$ in

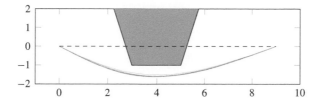

Figure 4.6: A 2D mobile robot motion planning problem.

Table 4.1

Comparison among different algorithms in C++.

Horizon	Iterations			Total Time (ms)			Time Per Iteration (ms)		
	CFS	ITP	SQP	CFS	ITP	SQP	CFS	ITP	SQP
10	5	42	86	4.74	8.53	201.41	0.95	0.20	2.34
20	4	46	366	4.80	17.37	1108.02	1.20	0.38	3.03
30	7	61	734	8.45	32.84	3164.40	1.21	0.54	4.31
40	4	87	1534	6.57	61.16	6717.45	1.64	0.70	4.38
50	4	139	1907	7.37	199.96	9633.93	1.84	1.44	5.05
60	4	140	3130	8.58	329.75	17123.95	2.14	2.36	5.47
70	4	165	2131	8.78	309.09	12546.80	2.19	1.87	5.89
80	4	169		9.96	307.55		2.49	1.82	
90	4	223		11.01	838.63		2.75	3.76	
100	4	253		11.80	1500.96		2.95	5.93	

Note: SQP exceeds maximum time limit for planning horizon longer than 70.

the convex feasible set. CFS scales much better than ITP and SQP, as the computation time and time per iteration in CFS goes up almost linearly with respect to number of steps.

4.4 OPTIMIZATION-BASED SPEED PROFILE PLANNING

Although CFS has good scalability, decoupling temporal planning from spatiotemporal planning is beneficial to further reduce the computational complexity. The objective of temporal planning is to obtain a speed profile for a given path.

As introduced in section 4.2, a speed profile is a one-to-one mapping between the time domain and the distance domain of the path. Optimization can be performed either over distance as shown in Fig. 4.7b or over time as shown in Fig. 4.7c. Optimization over distance requires an analytical parameterization of the path as discussed in [77, 78, 115, 174, 227]. However, globally continuous analytical parameterization of a complicated path is difficult, which usually requires approximation. For a curvy path, the approximation may introduce infeasibility, e.g., curvature exceeding the ve-

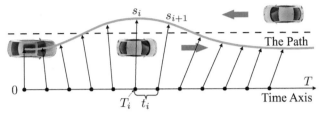

(a) Scenario: Overtake a slow front vehicle using the opposite lane.

(b) Optimization over station. For each sampled time step T_i, we need to find a corresponding station s_i on the path.

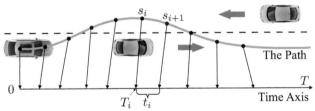

(c) Optimization over time. For each sampled station s_i, we need to find a corresponding time stamp T_i on the time axis.

Figure 4.7: Two optimization schemes to obtain a speed profile.

hicle's kinematic limits. Moreover, the complexity of the optimization problem may increase when complicated expressions of the parameterization enter the objective function. On the other hand, if we optimize over time as discussed in [220], only a sequence of way points is needed instead of a continuous parameterization of the path. This section focuses on speed profile planning via temporal optimization which optimizes the time stamps for all waypoints along a given path.

4.4.1 PROBLEM FORMULATION

Speed Profile and $s - T$ graph

Suppose a path $\mathbf{p}$ is given. There is no speed information in the path $\mathbf{p}$. In order to determine the speed at each station, we introduce an $s - T$ graph as shown in Fig. 4.8a. On the graph, the vertical axis is the one dimensional parameterization of the path and the horizontal axis is the time axis. A speed profile $\mathcal{V}$ for a path $\mathbf{p}$

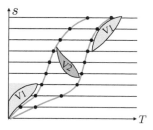
(a) Multiple topological speed profiles.

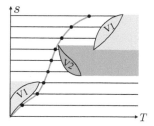
(b) One topological speed profiles.

Figure 4.8: Illustration of speed profile on $s - T$ graph. The scenario corresponds to Fig. 4.7a. The shaded area represents the moments that another road participant is occupying some parts of the path so that the ego vehicle cannot enter.

is a monotone curve on the $s - T$ graph as shown in Fig. 4.8a. Since the curve is monotone, there are two ways to obtain the speed profile. The first way is to find a mapping from the T axis to the s axis, e.g., fix a sequence of time steps $\{T_i\}$ and find the desired s_i for each T_i as shown in Fig. 4.7b. The second way is to find a mapping from the s axis to the T axis, e.g., fix a sequence of sampled stations $\{s_i\}$ and find the desired T_i for each s_i as shown in Fig. 4.7c. In either case, the speed profile is represented by the curve $\mathcal{V} = \{(T_i, s_i)\}_i$ in the $s - T$ graph.

Computing Speed, Acceleration and Jerk

Consider a speed profile $\mathcal{V} = \{(T_i, s_i)\}_i$. Denote $p_i := \mathbf{p}(s_i)$ and the time interval between p_i and p_{i+1} as $t_i = T_{i+1} - T_i$. When the time interval t_i and the distance between p_i and p_{i+1} are not too large, the velocity v_i, acceleration a_i and jerk j_i at p_i can be approximated by

$$v_i = \frac{p_{i+1} - p_i}{t_i}, \ a_i = \frac{2(v_i - v_{i-1})}{t_i + t_{i-1}}, \ j_i = \frac{3(a_i - a_{i-1})}{t_i + t_{i-1} + t_{i-2}} \tag{4.12}$$

Denote the heading of the vehicle at point p_i as $\theta_i := \arctan(\dot{\mathbf{p}}(s_i))$. The longitudinal and lateral directions at p_i are denoted as $\tau(\theta_i) := [\cos\theta_i, \sin\theta_i]$ and $\eta(\theta_i) := [\sin\theta_i, -\cos\theta_i]$ respectively. Then the longitudinal and lateral velocity, acceleration and jerk are defined as

$$v_i^\tau = v_i \cdot \tau(\theta_i), v_i^\eta = v_i \cdot \eta(\theta_i) \tag{4.13a}$$

$$a_i^\tau = a_i \cdot \tau(\theta_i), a_i^\eta = a_i \cdot \eta(\theta_i) \tag{4.13b}$$

$$j_i^\tau = j_i \cdot \tau(\theta_i), j_i^\eta = j_i \cdot \eta(\theta_i) \tag{4.13c}$$

Note that both p_i and θ_i depend on station s_i. If we optimize over distance, an analytical expression of $\mathbf{p}(s)$ is indispensable, which will also increase the complexity of the problem. On the other hand, if we optimize over time, s_i is predefined and p_i and θ_i are fixed for all i. The analytical expression of $\mathbf{p}(s)$ is not necessary. Moreover, in a lot of cases, the paths are represented by a sequence of points instead of an analytical expression. Hence temporal optimization is desired.

Speed Profile Planning via Temporal Optimization

As discussed earlier, given a path that is represented by a list of points $\{p_i\}_1^{h+1}$, we need optimize the time stamps T_i's for each point along the path. The safety constraints for the optimization problem is illustrated in the $s-T$ graph in Fig. 4.8a. The shaded area represents the moments that another road participant is occupying some parts of the path so that the ego vehicle cannot enter. The constraints in Fig. 4.8a are generated regarding the case shown in Fig. 4.7a where the slow front vehicle is denoted V1 and the vehicle in the opposite lane is denoted V2. As the ego vehicle can choose to enter the conflict zone before or after another road participant, there are more than one topological speed profile (or homotopy class) as shown in Fig. 4.8a. The red trajectory corresponds to the case that the ego vehicle overtook V1 before V2 passed by, while the green trajectory corresponds to the case that the ego vehicle overtook V1 after V2 went away. There is one unique local optimum for each topological speed profile (or homotopy class). It is assumed that the choice of a certain homotopy class is determined by a high level planner. In temporal optimization, the speed profile planner only tries to find the local optimum inside the chosen homotopy class as shown in Fig. 4.8b. By choosing one homotopy class, the high level planner would specify an interval constraint $[T_i^{min}, T_i^{max}]$ for the time stamp T_i at each point p_i,.

The temporal optimization problem is formulated as

$$\min_{\mathbf{t}} J_1(\mathbf{t}) + J_2(\mathbf{u}), \tag{4.14a}$$

$$s.t. \ T_i \in [T_i^{min}, T_i^{max}], \tag{4.14b}$$

$$|u_i| \leq \bar{u}, \tag{4.14c}$$

$$\mathcal{F}(\mathbf{t}, \mathbf{u}) = 0. \tag{4.14d}$$

where $\mathbf{t} := [t_1, \cdots, t_h]$, $\mathbf{u} := [u_1, \cdots, u_h]$, $u_i := [a_i^\tau, a_i^\eta, j_i^\tau, j_i^\eta, v^r - v_i^\tau]$,[1] and v^r is the reference speed. For simplicity, let $u_i^j \in \mathbb{R}$ be the j-th entry in u_i. Equation (4.14a) is the cost function, which penalizes cycle time for efficiency in J_1 and speed, acceleration and jerk for smoothness of the speed profile in J_2. J_1 and J_2 are convex. Moreover, J_2 is symmetric with respect to 0. Equation (4.14b) is the safety constraint to avoid dynamic obstacles. Equation (4.14c) is the dynamic constraint on the speed, acceleration and jerk, which is assumed to be a box constraint, i.e., a constraint that has symmetric lower and upper bounds. Equation (4.14d) encodes the relationship

---

[1] u_i may contain other parameters depending on the objective of the problem.

between the time stamps and the speed profile in (4.12) and (4.13). Indeed, (4.14d) can be written as a sequence of affine equations,

$$f_i^j(\mathbf{t}) + h_i^j(\mathbf{t})u_i^j = 0, \forall i = 1,\ldots,h, \forall j = 1,\ldots,5, \qquad (4.15)$$

where

$$
\begin{aligned}
f_i^1(\mathbf{t}) &= 2\left[t_i dp_{i-1}^{\tau_i} - t_{i-1} dp_i^{\tau_i}\right] \\
h_i^1(\mathbf{t}) &= t_i t_{i-1}(t_i + t_{i-1}) \\
f_i^2(\mathbf{t}) &= 2\left[t_i dp_{i-1}^{\eta_i} - t_{i-1} dp_i^{\eta_i}\right] \\
h_i^2(\mathbf{t}) &= t_i t_{i-1}(t_i + t_{i-1}) \\
f_i^3(\mathbf{t}) &= 6[(t_{i-1}+t_{i-2})t_{i-1}t_{i-2}dp_i^{\tau_i} - (t_i + 2t_{i-1} + t_{i-2})t_i t_{i-2} dp_{i-1}^{\tau_i} \\
&\quad + (t_i + t_{i-1})t_i t_{i-1} dp_{i-2}^{\tau_i}] \\
h_i^3(\mathbf{t}) &= t_i t_{i-1} t_{i-2}(t_i + t_{i-1})(t_{i-1} + t_{i-2})(t_i + t_{i-1} + t_{i-2}) \\
f_i^4(\mathbf{t}) &= 6[(t_{i-1}+t_{i-2})t_{i-1}t_{i-2}dp_i^{\eta_i} - (t_i + 2t_{i-1} + t_{i-2})t_i t_{i-2} dp_{i-1}^{\eta_i} \\
&\quad + (t_i + t_{i-1})t_i t_{i-1} dp_{i-2}^{\eta_i}] \\
h_i^4(\mathbf{t}) &= t_i t_{i-1} t_{i-2}(t_i + t_{i-1})(t_{i-1} + t_{i-2})(t_i + t_{i-1} + t_{i-2}) \\
f_i^5(\mathbf{t}) &= v^r t_i - dp_i^{\tau} \\
h_i^5(\mathbf{t}) &= t_i
\end{aligned}
$$

where $dp_i = p_{i+1} - p_i$, $dp_i^{\tau_j} = dp_i \cdot \tau(\theta_j)$, and $dp_i^{\eta_j} = dp_i \cdot \eta(\theta_j)$. By definition, $h_i^j \geq 0$ for all i and j. This property will be exploited to relax the problem. In order to compute the acceleration and jerk at $i = 1, 2$, we need to manually specify the constants p_0, p_{-1}, t_0 and t_{-1} according to the initial velocity v_0 and the initial acceleration $a_0 = 0$.

4.4.2   QUADRATIC APPROXIMATION

To solve (4.14) efficiently, we approximate it as a quadratic program using the slack convex feasible set algorithm (SCFS) [136].[2] SCFS is designed to convexify problems with non-linear equality constraints. The idea is to (1) relax the non-linear equality constraints to a set of non degenerating nonlinear inequality constraints using slack variables by exploiting the symmetry of the problem, and (2) approximate the relaxed problem as a quadratic program by CFS. A set of non-linear inequality constraints are non degenerating if they do not imply any nonlinear equality constraint.

It is assumed that J_1 and J_2 are quadratic. Since (4.14) is symmetric with respect to $\mathbf{u}$ in the cost function (4.14a) and in the dynamic constraints (4.14c), we introduce

---

[2]There are other methods [232] to convexify the speed profile planning problem.

a slack variable $\mathbf{y} \geq |\mathbf{u}|$. Then (4.14) can be relaxed as

$$\min_{\mathbf{t},\mathbf{y}} J_1(\mathbf{t}) + J_2(\mathbf{y}) \tag{4.16a}$$

$$s.t.\ A\mathbf{t} \leq b, y_i \leq \bar{u} \tag{4.16b}$$

$$f_i^j(\mathbf{t}) + h_i^j(\mathbf{t})y_i^j \geq 0, f_i^j(\mathbf{t}) - h_i^j(\mathbf{t})y_i^j \leq 0 \tag{4.16c}$$

where

$$A = \begin{bmatrix} 1 & 0 & \cdots & 0 \\ 1 & 1 & \ddots & 0 \\ \vdots & \ddots & \ddots & 0 \\ 1 & \cdots & 1 & 1 \\ -1 & 0 & \cdots & 0 \\ -1 & -1 & \ddots & 0 \\ \vdots & \ddots & \ddots & 0 \\ -1 & \cdots & -1 & -1 \end{bmatrix}, b = \begin{bmatrix} T_1^{max} \\ \vdots \\ T_h^{max} \\ -T_1^{min} \\ \vdots \\ -T_h^{min} \end{bmatrix}$$

It is proven in [136] that (4.16) is equivalent to (4.14) in the sense that: if $(\mathbf{t}^o, \mathbf{y}^o)$ is a local optimum of (4.16), then $\mathbf{t}^o$ is a local optimum of (4.14); and if $\mathbf{t}^o$ is a local optimum of (4.14), then $(\mathbf{t}^o, \mathbf{y}^o)$ is a local optimum of (4.16) with $(y_i^j)^o = |[h_i^j(\mathbf{t}^o)]^{-1} f_i^j(\mathbf{t}^o)|$. The intuition is that the equality constraints in (4.14) hold in the optima of (4.16), i.e., $\mathbf{y} = |\mathbf{u}|$, and either $f_i^j(\mathbf{t}) + h_i^j(\mathbf{t})y_i^j = 0$ or $f_i^j(\mathbf{t}) - h_i^j(\mathbf{t})y_i^j = 0$. Hence solving the relaxed problem is equivalent to solving the original problem. Since the relaxed problem does not have any equality constraint (i.e., (4.16c) does not degenerate to an equality constraint), it is easier to quadratify.

The relaxed problem is solved iteratively. Suppose at the kth iteration, we have the solution $\mathbf{t}^{(k)}$ and $\mathbf{y}^{(k)}$. Then at the $(k+1)$th iteration, (4.16c) is linearized with respect to $\mathbf{t}^{(k)}$ and $\mathbf{y}^{(k)}$, and the following approximated quadratic program is formulated,

$$\min_{\mathbf{t},\mathbf{y}} J_1(\mathbf{t}) + J_2(\mathbf{y}) \tag{4.17a}$$

$$s.t.\ A\mathbf{t} \leq b, y_i \leq \bar{u} \tag{4.17b}$$

$$\left[\nabla_t f_i^j(\mathbf{t}^{(k)}) + \nabla_t h_i^j(\mathbf{t}^{(k)})(y_i^j)^{(k)}\right](\mathbf{t} - \mathbf{t}^{(k)}) + h_i^j(\mathbf{t}^{(k)})y_i^j + f_i^j(\mathbf{t}^{(k)}) \geq 0 \tag{4.17c}$$

$$\left[\nabla_t f_i^j(\mathbf{t}^{(k)}) - \nabla_t h_i^j(\mathbf{t}^{(k)})(y_i^j)^{(k)}\right](\mathbf{t} - \mathbf{t}^{(k)}) - h_i^j(\mathbf{t}^{(k)})y_i^j + f_i^j(\mathbf{t}^{(k)}) \leq 0 \tag{4.17d}$$

Solving the above quadratic program, we obtain $\mathbf{t}^o$ and $\mathbf{y}^o$. Define $\mathbf{t}^{(k+1)} = \mathbf{t}^o$ and $(y_i^j)^{(k+1)} = |[h_i^j(\mathbf{t}^o)]^{-1} f_i^j(\mathbf{t}^o)|$. Then we iterate until the solutions converge, i.e., $\|\mathbf{t}^{(k+1)} - \mathbf{t}^{(k)}\| \leq \varepsilon$ for ε small. $\mathbf{t}^{(0)}$ and $\mathbf{y}^{(0)}$ are initialized as

$$t_i^{(0)} = \frac{dp_i^{\tau_i}}{v^*}, (y_i^j)^{(0)} = |[h_i^j(\mathbf{t}^{(0)})]^{-1} f_i^j(\mathbf{t}^{(0)})| \tag{4.18}$$

where v^* can be the reference velocity v^r or the initial speed v_0^τ.

4.4.3 EXAMPLE: VEHICLE SPEED PROFILE PLANNING

Autonomous driving is widely viewed as a promising technology to revolutionize today's transportation system. However, it is still challenging to plan collision-free, time-efficient and comfortable trajectories for an automated vehicle in dynamic environments such as urban roads, since the vehicle needs to interact with other road participants. For example, how to overtake a slow front vehicle safely using the opposite lane as shown in Fig. 4.7a?

For vehicles, the path $\mathbf{p}$ in $\mathbb{R}^2$ concerns with the position of the center of the rear axle. Regarding the driving quality, a^τ and j^τ should be minimized for longitudinal comfort, while a^η and j^η should be minimized for lateral comfort. Moreover, a reference speed v^r should be tracked for time efficiency. Hence, the cost function in (4.14a) is designed to be

$$J_1(\mathbf{t}) = 0, J_2(\mathbf{u}) = w_1 \sum \|a_i^\tau\|^2 + w_2 \sum \|a_i^\eta\|^2 + w_3 \sum \|j_i^\tau\|^2, \qquad (4.19)$$

where $w_1 = 1$ and $w_2 = w_3 = w_4 = w_5 = 10$ are weights. The dynamic constraint in (4.14c) is designed to be

$$|a_i^\tau| \leq \bar{a}, |a_i^\eta| \leq \bar{a}, \qquad (4.20)$$

where $\bar{a} := 2.5\,\mathrm{m\,s^{-2}}$ for passenger comfort.

The performance of the speed profile planning will be illustrated in the overtaking scenario in Fig. 4.7a. The simulations were run in MATLAB® on a MacBook of 2.3 GHz using Intel Core i7. In SCFS, (4.17) for each iteration was solved using the `quadprog` function. For comparison, (4.14) was also solved using SQP in the `fmincon` function. SCFS and SQP terminated if the step size (difference between the two consecutive solutions) was less than 10^{-6}. For better illustration, the speed profiles after every iteration are shown in grayscale (the later in the iteration, the darker) together with the optimal speed profile in all three cases.

The scenario is shown in Fig. 4.7a where the automated vehicle wants to overtake the slow front vehicle using the opposite lane. The reference speed is $v_r = 11\,\mathrm{m\,s^{-1}}$. The vehicle's initial speed is $v_0^\tau = 10\,\mathrm{m\,s^{-1}}$ and $v_0^\eta = 0\,\mathrm{m\,s^{-1}}$. $\mathbf{t}^{(0)}$ is initialized using v_0^τ. The path is sampled every 2 m and 34 points are chosen. SCFS converges at iteration 5 with computation time 0.308 s. The optimal speed, acceleration and jerk profiles are shown in Fig. 4.9a. The horizontal axis in the plots represents the traveling distance along the lane. The corresponding time stamps for all stations are shown in Fig. 4.9b. Snapshots are also shown in Fig. 4.9b where the gray rectangle represents the ego vehicle and the yellow rectangles are the surrounding vehicles. The ego vehicle slowed down first to keep a safe headway from the front vehicle V1. When it changed to the adjacent lane, it speeded up to overtake V1. Before the vehicle V2 in the opposite direction came, the ego vehicle went back to its lane. The optimal speed profile is on the boundary of the safety constraint as shown in Fig. 4.9b and on the boundary of the feasibility constraint as shown in the acceleration profile in Fig. 4.9a. For comparison, SQP converges to the same optimum after 48 iterations with computation time 28.513 s.

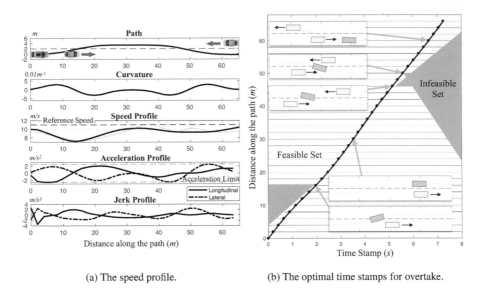

(a) The speed profile.   (b) The optimal time stamps for overtake.

Figure 4.9: Speed profile planning for overtake.

4.5 OPTIMIZATION-BASED LAYERED PLANNING

This section discusses a framework for layered planning as shown in Fig. 4.10. The framework has two layers, a trajectory planning layer and a temporal optimization layer. Each layer deals with safety and efficiency respectively. Trajectory planning is to plan a fixed-horizon collision-free trajectory from a given reference trajectory or several way points as discussed in section 4.3. The temporal optimization is to minimize the cycle time over the planned trajectory as discussed in section 4.4. Trajectory planning can also be substituted by path planning.

We illustrate the performance of the layered planning on a six degree of freedom robot arm. The problem is illustrated in Fig. 4.11a where the robot arm needs to go to the five way points (red dots) in sequence. The reference paths (shown in blue) between two consecutive way points are chosen by linear interpolation between the way points. Paths 1, 2, and 5 are collision-free, while Paths 3 and 4 are occupied by obstacles. The objective is to optimize both the quality of the path and the cycle time to traverse the paths. There is a natural decomposition to first optimize the path and then reduce cycle time.

1. Trajectory planning
   Trajectory planner solves (4.8). The planning horizon is $h = 95$. We compare the performance of CFS and SQP in MATLAB® on a Windows desktop with an Intel Core i5 CPU. Table 4.2 shows the computation time of CFS and SQP. On the collision-free paths, there was not significant dif-

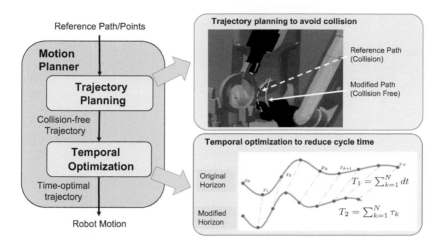

Figure 4.10: A framework for layered planning.

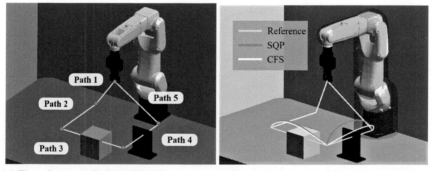

(a) The reference path that robot plans to move, where the red dots are the way points.

(b) The simulation result of the trajectory planning with SQP and CFS.

Figure 4.11: The reference path and planned path.

ference between the two algorithms. On the blocked paths, CFS computes much faster and requires fewer iterations than SQP. The planned trajectories are shown in Fig. 4.11b, where the red line and the yellow line represents trajectories found by SQP and CFS respectively. Although these algorithms converged to different solutions, both trajectories are locally optimal.

2. Speed profile planning
   Speed profile planner solves (4.14) for the planned trajectory. We penalize the cycle time in J_1, i.e., the total time to traverse the paths. The expected outcome is illustrated in Fig. 4.12, where the trajectory is speeded up by adjusting the time stamps at each way point. The problem is solved using

Table 4.2

Comparison between SQP and CFS on trajectory planning.

| | Sequential Quadratic Programming (SQP) | | Convex Feasible Set Algorithm (CFS) | |
	Computation Time	Iterations	Computation Time	Iterations
Path 1	0.0043	0	0.0033	0
Path 2	0.0031	0	0.0033	0
Path 3	46.8098	94	0.8938	2
Path 4	48.8767	96	0.8397	5
Path 5	0.0032	0	0.0031	0
Total	95.725	190	1.7434	7

Note: Unit in second (s). Total planning horizon is 95 step.

SCFS with respect to the trajectory obtained by CFS in the trajectory planner. The hardware set up remains the same. The computation time for it is shown in Table 4.3, where the average computation time per path is 31 ms. The cycle time is reduced from 15 s to 6.63 s without much increase in the computation load.

The example demonstrates both the motion efficiency and computation efficiency. A comprehensive discussion of the layer planning framework can be found in [117].

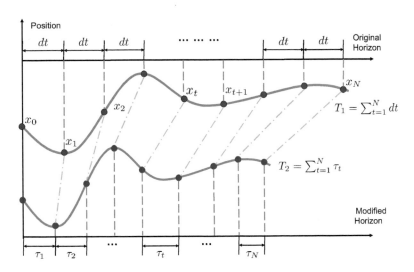

Figure 4.12: Illustration of the temporal optimization.

Table 4.3

Computation time for layered planning.

	Trajectory Planning	Temporal Optimization
Path 1	0.0033	0.0279
Path 2	0.0033	0.0272
Path 3	0.8938	0.0272
Path 4	0.8397	0.0485
Path 5	0.0031	0.0252
Total	1.7434	0.1561

Note: Unit in second (s). Total planning horizon is 95 steps.

4.6 CONCLUSION

In this chapter, we discussed the optimization-based methods for global motion planning to make robots efficient. In particular, we discussed optimization-based trajectory planning, optimization-based speed profile planning, and optimization-based layered planning. CFS and SCFS were adopted to transform the non-convex optimization problems into quadratic programs. Various examples illustrate the effectiveness of the methods.

5 Imitation: Mimicking Human Behavior

5.1 OVERVIEW

Human behavior is a key factor in human-robot interactions. It is important for a robot to understand human behavior in order to (1) correctly predict the human motion in the future so as to safely and efficiently plan its own motion, and (2) learn skills to finish various tasks from human demonstration. In the first case, the robot only mentally simulates human behavior to make the prediction. In the second case, the robot needs to physically execute the learned skill models from human demonstration.

As discussed in section 1.4.3, the major limitation of existing imitation approaches for prediction is that the learned model is not individualized, which may not be able to track individual differences and time varying behaviors. The major limitation of existing imitation approaches for skill learning is the lack of physical interpretation and stability proof, which is crucial for implementation on industrial robots.

This chapter discusses methods to improve prediction and skill learning through imitation. Section 5.2 discusses imitation for prediction through offline behavior classification and online model adaptation. Section 5.3 discusses imitation for action (or skill learning) in a model-free framework.

5.2 IMITATION FOR PREDICTION

Imitation for prediction is a cognition skill. The dynamic model of the environment, especially the closed-loop dynamics of human agents, needs to be identified. It is a part of the learning process and can be performed either in the training phase (offline) or in the execution phase (online). Nonetheless, the learning algorithms should be specified in the design phase. This section discusses the algorithms according to the structure described in section 2.2.1 and shown in Fig. 2.8.

As pointed out in section 1.3.2, there are multiple ways to model human behavior. Among those models, we choose reactive models to describe the behavior of other agents. Denote the state of the robot that we design for as x_R and the state of the other agents as x_H. Ignoring the noise term, the reactive model (2.8) can be rewritten in discrete time as

$$x_H(k+1) = F(x(k)) \tag{5.1}$$

There are many methods to identify (5.1), for example, Gaussian mixture model (GMM) [148] or neural networks [178]. The structure in Fig. 2.8 is adopted, which corresponds to the following equations mathematically,

$$x_H(k+1) \;=\; F_{\delta(k)}(x(k)) \tag{5.2}$$

$$F_m(x(k)) \;=\; \sum_j \theta_{m,j} f_{m,j}(x(k)) k_m \tag{5.3}$$

where $\delta(k) \in \{1, \ldots, M\}$ is a discrete classification function. M is the number of models. F_m is the closed loop dynamics in model m, which is a linear combination of several features $f_{m,j}(x)$. $\theta_{m,j} \in \mathbb{R}$ are coefficients that can be adapted online. Mathematically, F_m can be understood as a group of basis for function F. The features should either be designed or identified in the training phase. In [45], we use a neural network to approximate F_m. The network functions as a feature extractor. The nodes in the last hidden layer in the network can be regarded as features $f_{m,j}$. The weights from the last hidden layer to the output layer are the parameters $\theta_{m,j}$. These parameters are adaptable online as will be discussed in section 5.2.2.

To obtain a high-fidelity prediction model (5.2), we first need to do model selection in (5.2) and then model adaptation in (5.3).

5.2.1 CLASSIFICATION OF BEHAVIORS

There are multiple approaches for model classification and selection. This section introduces an approach based on Hidden Markov Model (HMM) [41].

Dynamic Classification using Hidden Markov Model

Under HMM, the classification function $\delta(k)$ is regarded as the hidden variable that needs to be inferred from the observation $\pi_R(k)$. There are two important relationships in a HMM:

1. How the current classification is affected by previous classification (or how the intentions of agents change dynamically)?
2. How the current classification is affected by current observation (or how the intentions of agents are revealed in data)?

The first corresponds to a dynamic model, which encodes the following relationship

$$\mathbb{P}(\delta(k) = m \mid \delta(k-1) = n), \forall m, n \tag{5.4}$$

The second corresponds to a measurement model, which encodes the following relationship

$$\mathbb{P}(y_R(k) \mid \delta(k) = m), \forall m \tag{5.5}$$

Then the conditional probability distribution of $\delta(k)$ given $\pi_R(k)$ is

$$\mathbb{P}(\delta(k) = m \mid \pi_R(k)) \tag{5.6}$$
$$\propto \quad \mathbb{P}(\delta(k) = m, y_R(1), \cdots, y_R(k))$$
$$\propto \quad \mathbb{P}(y_R(k) \mid \delta(k) = m) \mathbb{P}(\delta(k) = m \mid \pi_R(k-1))$$
$$\propto \quad \mathbb{P}(y_R(k) \mid \delta(k) = m) \sum_n \mathbb{P}(\delta(k) = m \mid \delta(k-1) = n) \mathbb{P}(\delta(k-1) = n \mid \pi_R(k-1))$$

which depends on the conditional probability distribution $\mathbb{P}(\delta(k-1) = n \mid \pi_R(k-1))$ in the last time step, the dynamic model (5.4) and the measurement model (5.5). Then the estimate of $\delta(k)$ is chosen according to maximum likelihood.

The dynamic model (5.4) and the measurement model (5.5) can either be designed in the design phase or be learned in the training phase. The training data can be obtained through: (1) real world experiments; (2) virtual reality-based simulations as discussed in section 1.5.2. HMM requires supervised learning. Hence the data needs to come with label $\delta(k)$.

Example: Behavior Classification of Surrounding Vehicles

In autonomous driving, it is important to correctly detect the intention of the surrounding vehicles. In this example, three intended behaviors are considered:

Behavior 1 ($\mathbb{B}_1$): Lane following
Behavior 2 ($\mathbb{B}_2$): Lane changing to the left
Behavior 3 ($\mathbb{B}_3$): Lane changing to the right

The transition model among the three behaviors is designed to be

$$A = \begin{bmatrix} 0.6 & 0.5 & 0.5 \\ 0.2 & 0.5 & 0 \\ 0.2 & 0 & 0.5 \end{bmatrix} \tag{5.7}$$

where the mth row and nth column entry is $A_{m,n} = \mathbb{P}(\delta(k) = m \mid \delta(k-1) = n)$.

The measurement model for a vehicle is designed according to the following three features: (1) its lateral velocity v^η, (2) its lateral deviation from the center of its current lane d ($d > 0$ if the deviation is to the left), and (3) an indicator c that shows whether the vehicle is crossing the boundary of two lanes ($c = 1$ if true). Then the measurement model is

$$\mathbb{P}(y_R(k) \mid \delta(k) = 1) \quad = \quad (1-c)e^{-(v^\eta + d)^2} \tag{5.8}$$
$$\mathbb{P}(y_R(k) \mid \delta(k) = 2) \quad = \quad c\mathbf{1}(d \geq 0)(1 - e^{-(v^\eta + d)^2}) \tag{5.9}$$
$$\mathbb{P}(y_R(k) \mid \delta(k) = 3) \quad = \quad c\mathbf{1}(d < 0)(1 - e^{-(v^\eta + d)^2}) \tag{5.10}$$

where $\mathbf{1}(\cdot)$ is an indicator function, which is 1 if the condition in the parenthesis is true, and 0 otherwise. Given the transition model and measurement model, the distribution of $\delta(k)$ follows from (5.6). The simulation result is shown in Fig. 5.1, which correctly predicts the lane change behavior before the yellow vehicle crossed the lane boundary.

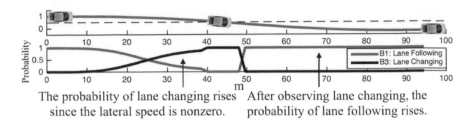

The probability of lane changing rises After observing lane changing, the
since the lateral speed is nonzero.   probability of lane following rises.

Figure 5.1: Behavior classification for a vehicle in a two-lane case.

5.2.2 ADAPTATION OF BEHAVIOR MODELS

Online Learning Using Parameter Adaptation

For a selected model m, the learning objective is to approximate the function
$F_m(x(k))$ such that the estimate $\hat{x}_H(k+1) = \hat{F}_m(x(k))$ minimizes the expected prediction error, i.e.,

$$\hat{x}_H(k+1) = \arg\min_a \mathbb{E}_{x_H(k+1)}(\|x_H(k+1) - a\|^2) \tag{5.11}$$

Define the *a priori* estimate of x_H as the optimal solution of (5.11) given information
up to the last time step, and the *a posteriori* estimate of x_H as the optimal solution
of (5.11) given information up to the current time step. Similarly, we can define *a
priori* and *a posteriori* estimation error and mean squared estimation error (MSEE).
The notations are listed in Table 5.1.

Table 5.1

The notations in state estimation and prediction.

	State Estimate	Estimation Error	MSEE
a posteriori	$\hat{x}_H(k\|k)$	$\tilde{x}_H(k\|k) = x_j(k) - \hat{x}_j(k\|k)$	$X_H(k\|k)$
a priori	$\hat{x}_H(k+1\|k)$	$\tilde{x}_H(k+1\|k) = x_j(k+1) - \hat{x}_j(k+1\|k)$	$X_H(k+1\|k)$

It is assumed that the robot can directly measure the human state. The robot's
measurement of the human is denoted as y_R^H, which satisfies that

$$y_R^H(k) = x_H(k) + v_R^H(k) \tag{5.12}$$

where $v_R^H(k)$ is the measurement noise assumed to be zero-mean, Gaussian and
white.

Rewriting $F_m(x(k))$ in (5.3) and ignoring m, we have the following dynamics

$$x_H(k+1) = \Phi(x(k))\vartheta^\mathsf{T} + w(k) \tag{5.13}$$

where $\Phi(\cdot) = [f_1(\cdot), \dots, f_j(\cdot), \dots]$, $\vartheta = [\theta_1, \dots, \theta_j, \dots]$ and w is a noise term assumed to be zero-mean, Gaussian, white and with covariance W.

Equations (5.13) and (5.12) form a non-linear Gaussian system with unknown parameters. A recursive least square parameter adaptation algorithm (RLS-PAA) [73] is developed to identify the system online so that the prediction $\hat{x}_H(k+1|k)$ minimizes the expected prediction error (5.11) given information up to time step k.

Define $\hat{\vartheta}(k)$ to be the estimates of the coefficients given information up to the time step k. The state and parameter can be estimated following the procedures below:

1. State Estimation

   At $(k+1)$th time step, $\hat{x}_H$ is first updated according to the closed loop dynamics in (5.14). Then the measurement information is incorporated in the *a posteriori* estimate in (5.15). A constant update gain $\alpha \in (0, 1)$ is chosen to ensure that the measurement information is always incorporated.

$$\hat{x}_H(k+1|k) = \Phi(\hat{x}(k|k))\hat{\vartheta}(k)^\mathsf{T}, \tag{5.14}$$

$$\hat{x}_H(k+1|k+1) = (1-\alpha)\hat{x}_H(k+1|k) + \alpha y_R^H(k+1), \tag{5.15}$$

   where $\hat{x}(k|k) = [\hat{x}_H(k|k)^\mathsf{T}, \hat{x}_R(k|k)^\mathsf{T}]^\mathsf{T}$, and $\hat{x}_R(k|k)$ is the estimate of the state of the robot itself, which can be obtained using Kalman Filter [183].

2. Parameter Estimation

   The coefficients are estimated using RLS-PAA [73]:

$$\hat{\vartheta}(k+1) = \hat{\vartheta}(k) + (\hat{x}_H(k+1|k+1) - \hat{x}_H(k+1|k))^\mathsf{T} \Phi(\hat{x}(k|k))F(k+1) \tag{5.16}$$

   where F is the learning gain such that

$$F(k+1) = \frac{1}{\lambda}\left[F(k) - \frac{F(k)\Phi(\hat{x}(k|k))^\mathsf{T}\Phi(\hat{x}(k|k))F(k)^\mathsf{T}}{\lambda + \Phi(\hat{x}(k|k))F(k)\Phi(\hat{x}(k|k))}\right] \tag{5.17}$$

   where $\lambda \in (0, 1)$ is a forgetting factor.

Quantifying the Uncertainty

The motions of other agents can be predicted using the estimate $\hat{\vartheta}$ in the model (5.13). However, we also need to quantify the uncertainty associated with the prediction in terms of MSEE,

$$X_H(k+h|k) = E_{x_H(k+h)}(\|x_H(k+h) - \hat{x}_H(k+h|k)\|^2) \tag{5.18}$$

Assume that the human state can be perfectly measured, i.e., $v_R^H(k) = 0$ in (5.12). Let $\tilde{\vartheta}(k) = \vartheta(k) - \hat{\vartheta}(k)$ be the parameter estimation error. The *a priori* state estimation error is

$$\tilde{x}(k+1|k) = \Phi(x(k))\tilde{\vartheta}(k)^\mathsf{T} + w(k) \tag{5.19}$$

Since $\hat{\vartheta}(k)$ only contains information up to the $(k-1)$th time step, $\tilde{\vartheta}(k)$ is independent of $w(k)$. Thus the *a priori* MSEE is

$$X_{\tilde{x}\tilde{x}}(k+1|k) = E\left[\tilde{x}(k+1|k)\,\tilde{x}(k+1|k)^{\mathsf{T}}\right] = \Phi(k)X_{\tilde{\vartheta}\tilde{\vartheta}}(k)\,\Phi^{\mathsf{T}}(k)+W, \qquad (5.20)$$

where $X_{\tilde{\vartheta}\tilde{\vartheta}}(k) = E\left[\tilde{\vartheta}(k)^{\mathsf{T}}\,\tilde{\vartheta}(k)\right]$ is the mean squared parameter estimation error. The parameter estimation error is

$$\tilde{\vartheta}(k+1) = \tilde{\vartheta}(k) - F(k+1)\,\Phi^{\mathsf{T}}(k)\,\tilde{x}(k+1|k) + \Delta\vartheta(k) \qquad (5.21)$$

where $\Delta\vartheta(k) := \vartheta(k+1) - \vartheta(k)$ is the temporal difference of the parameter. Since the system is time varying, the estimated parameter is biased and the expectation of the error can be expressed as

$$
\begin{aligned}
E\left(\tilde{\vartheta}(k+1)\right) &= \left[I - F(k+1)\Phi^{\mathsf{T}}(k)\Phi(k)\right]E\left(\tilde{\vartheta}(k)\right) + \Delta\vartheta(k), \\
&= \sum_{n=0}^{k}\prod_{i=n+1}^{k}\left[I - F(i+1)\Phi^{\mathsf{T}}(i)\Phi(i)\right]\Delta\vartheta(n). \qquad (5.22)
\end{aligned}
$$

The second equality is by induction.

The mean squared parameter estimation error follows from (5.21) and (5.22):

$$
\begin{aligned}
&X_{\tilde{\vartheta}\tilde{\vartheta}}(k+1) \qquad\qquad\qquad\qquad\qquad\qquad\qquad\qquad\qquad\qquad (5.23)\\
=\; & F(k+1)\Phi^{\mathsf{T}}(k)X_{\tilde{x}\tilde{x}}(k+1|k)\Phi(k)F(k+1)\\
& -X_{\tilde{\vartheta}\tilde{\vartheta}}(k)\Phi^{\mathsf{T}}(k)\Phi(k)F(k+1) - F(k+1)\Phi^{\mathsf{T}}(k)\Phi(k)X_{\tilde{\vartheta}\tilde{\vartheta}}(k)\\
& +E\left[\tilde{\vartheta}(k+1)\right]\Delta\vartheta^{\mathsf{T}}(k) + \Delta\vartheta(k)E\left[\tilde{\vartheta}(k+1)\right]^{\mathsf{T}} - \Delta\vartheta(k)\Delta\vartheta(k)^{\mathsf{T}} + X_{\tilde{\vartheta}\tilde{\vartheta}}(k)
\end{aligned}
$$

Since $\Delta\vartheta(k)$ is unknown in (5.22) and (5.23), it is set to an average time varying rate in the implementation.

Example: Identification of a Linear Time Varying System

In this example, we illustrate the performance of the algorithms on a linear scalar system

$$x(k+1) = A(k)x(k) + B(k)u(k) + w(k) \qquad (5.24)$$

where x, u, A, and B are all scalars. A and B are unknown coefficients that need to be identified. B is constant, while A is time varying. $x(k)$ and $u(k)$ correspond to linear features in (5.13), which is directly measurable.

Figure 5.2 shows the simulation result of the proposed learning algorithm on a first order system with a noise covariance $W = 0.005^2$. A forgetting factor $\lambda = 0.98$ is used. The solid and dashed blue lines in the upper figure are $\hat{A}(k)$ and $A(k)$, while the solid and dashed green lines are $\hat{B}(k)$ and the constant parameter B respectively. As shown in the figure, the time varying parameter $A(k)$ is well approximated by

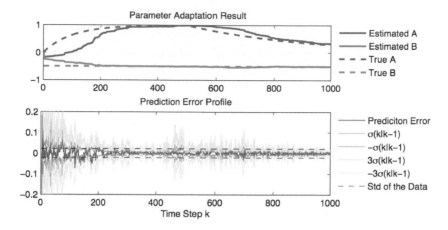

Figure 5.2: Identification of a linear time varying system.

$\hat{A}(k)$, while $\hat{B}(k)$ converges to B. In the lower figure, the blue curve is the one step prediction error $\tilde{x}(k|k-1)$. The green curves are the 3σ bound ($\sigma = \sqrt{X_{\tilde{x}\tilde{x}}(k|k-1)}$). The black dashed line is the statistical standard deviation (Std) of the data $\tilde{x}(k|k-1)$ from $k=1$ to $k=1000$. As shown in the figure, the 3σ value offers a good bound for the prediction errors as all measured errors lie between the green curves. Moreover, the MSEE is larger when the parameter is changing faster, which captures the time varying property of the system. On the other hand, the statistical standard deviation does not provide a good estimation of uncertainty in real time.

Note that although the method works well in practice, it is possible that it goes unstable for complicated dynamics. Theoretical analysis will be left for future work.

5.3   IMITATION FOR ACTION

A human skill can be regarded as an unknown control policy, which takes sensor inputs and generates action outputs accordingly. Model-free statistical approaches can be deployed to learn the skill by regressing the mapping between input and output. Among all predictors for regression, the question is: which would be a suitable model-free method for robotic learning?

We set up three criteria to select a proper predictor:

1. *Stability.* Since the learned controller might be embedded in the feedback control loop of the robotic system, it could influence the stability of the whole system. The stability conditions of the predictor should be explicitly formulated for analysis.

2. *Efficiency.* The computation capacity of robotic systems is usually limited in order to make the whole system cost effective, especially for industrial

robots. It is preferred that the predictor has a closed-form expression to be computed efficiently online.

3. *Interpretability*. Many prediction methods behave like black boxes. By extensive training, they could generate proper output, but have limited transparency or explainability on their internal dynamics. It is preferred that the predictor has an explicit physical interpretation.

Many common predictors, such as neural networks, logistic regression, K-nearest neighbors cannot meet all the above criteria. To meet the three criteria, we introduce the Gaussian mixture regression (GMR) [28] to learn the human skill. In the rest of this section, we will show that GMR has good efficiency, reasonable physical interpretations, and explicit stability conditions.

5.3.1   IMITATION LEARNING BY GAUSSIAN MIXTURE MODEL

For the ease of illustration, let's take the task of robotic assembly as an example. During the assembly of two workpieces, e.g., peg-hole insertion, robots need to actively adjust the insertion velocity between the two workpieces given the feedback of contact wrench between them. Humans are able to assemble workpieces together compliantly, robustly, and efficiently. However, it is still challenging for robots to do the same at the current stage of technology. Therefore it is demanded that the human assembly skills are transferred to robots. During an assembly demonstration, the input of the skill is the sensed wrench w, while the control output is the corrective assembly velocity $\dot{x}_c$ that human decides. The basic idea of GMR is to encode the skill behind human demonstration into a joint probability distribution $p(w,\dot{x}_c)$, and then use its conditional probability $p(\dot{x}_c|w)$ to retrieve the output $\dot{x}_c$, given the input w.

As shown in Fig. 5.3, the first step of GMR is to fit the joint probability distribution $p(w,\dot{x}_c)$ by the weighted sum of N Gaussian mixtures, each Gaussian with mean μ_i, covariance Σ_i and weight α_i

$$p(w,\dot{x}_c) = \sum_{i=1}^{N} \alpha^i p^i(w,\dot{x}_c) \tag{5.25}$$

$$= \sum_{i=1}^{N} \alpha^i \mathcal{N}\left(\begin{bmatrix} w \\ \dot{x}_c \end{bmatrix} \Big| \begin{bmatrix} \mu_w^i \\ \mu_{\dot{x}_c}^i \end{bmatrix}, \begin{bmatrix} \Sigma_w^i & \Sigma_{w\dot{x}_c}^i \\ \Sigma_{\dot{x}_c w}^i & \Sigma_{\dot{x}_c}^i \end{bmatrix} \right) \tag{5.26}$$

where $\sum_{i=1}^{N} \alpha^i = 1$.

Since each component $p^i(w,\dot{x}_c)$ is Gaussian, its conditional probability distribution $p^i(\dot{x}_c \mid w)$ is still Gaussian:

$$p^i(\dot{x}_c \mid w) = \mathcal{N}(\dot{x}_c \mid \mu_{\dot{x}_c|w}^i, \Sigma_{\dot{x}_c|w}^i) \tag{5.27}$$

where $\mu_{\dot{x}_c|w}^i$ is the conditional mean and $\Sigma_{\dot{x}_c|w}^i$ is the conditional variance. The overall conditional probability $p(\dot{x}_c \mid w)$ can be constructed as the weighted sum of $p^i(\dot{x}_c \mid w)$

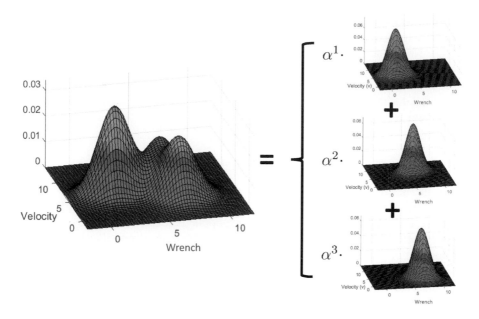

Figure 5.3: Estimate data distribution by Gaussian mixture. For the ease of visualization, both velocity and wrench are one-dimensional here. The distribution density of the velocity and wrench data collected from human demonstration is shown on the left. This density map is then fitted by superposition of several weighted Gaussian distributions with different means and covariance.

for all i. The weight for each component indicates the probability that w belongs to that Gaussian component,

$$p(\dot{x}_c \mid w) = \sum_{i=1}^{N} \frac{\alpha^i \mathcal{N}(w \mid \mu_w^i, \Sigma_w^i)}{\sum_{j=1}^{N} \alpha^j \mathcal{N}(w \mid \mu_w^j, \Sigma_w^j)} p^i(\dot{x}_c \mid w) \qquad (5.28)$$

The skill module is constructed such that given an input wrench w, it will generate an output corrective velocity $\dot{x}_c^*$ that maximizes $p(\dot{x}_c \mid w)$. The output corrective velocity $\dot{x}_c^*$ could be calculated from (5.29) directly. Since (5.29) is an explicit function, the this control policy can be implemented in real time with low computation load.

$$\dot{x}_c^* = \arg\max_{\dot{x}_c} p(\dot{x}_c \mid w)$$

$$= \sum_{i=1}^{N} \frac{\alpha^i \mathcal{N}(w \mid \mu_w^i, \Sigma_w^i)}{\sum_{j=1}^{N} \alpha^j \mathcal{N}(w \mid \mu_w^j, \Sigma_w^j)} \mu_{\dot{x}_c \mid w}^i$$

$$= \sum_{i=1}^{N} \frac{\alpha^i \mathcal{N}(w \mid \mu_w^i, \Sigma_w^i)}{\sum_{j=1}^{N} \alpha^j \mathcal{N}(w \mid \mu_w^j, \Sigma_w^j)} \left[\mu_{\dot{x}_c}^i + \Sigma_{\dot{x}_c w}^i (\Sigma_w^i)^{-1} (w - \mu_w^i) \right] \qquad (5.29)$$

In GMR, the Gaussian parameters $(\mu^i, \Sigma^i, \alpha^i)$ can be estimated iteratively by the expectation maximization (EM) algorithm [27] from the human demonstration data $(w, \dot{x}_c)$. Note that since the performance of parameter identification is sensitive to the initialization, the K-means clustering [85] is applied first on the dataset to get a good initialization. Specifically, $(w, \dot{x}_c)$ is clustered into N groups according to the Euclidean distance, and then each group is treated as an initial Gaussian component. The parameters $(\mu^i, \Sigma^i, \alpha^i)$ are calculated for each group i.

5.3.2   INTERPRETING GMR FROM MECHANICS, POINT OF VIEW

From the mechanics point of view, *admittance* denotes how much a structure resists motion when subjected to an external force [94]. To be specific, it is the ratio of the resulting velocity to the external force applied on a structure. The aforementioned GMR is trying to construct an appropriate *admittance controller* from human demonstration data. However, the statistic form of GMR in (5.29) does not provide insight to the dynamic feature of the admittance. In this section, we rewrite (5.29) to formulate a dynamic system, and give this GMR-based controller an explicit physical interpretation. To simplify the notation, we define:

$$h^i(w) = \frac{\alpha^i \mathcal{N}(w \mid \mu_w^i, \Sigma_w^i)}{\sum_{j=1}^N \alpha^j \mathcal{N}(w \mid \mu_w^j, \Sigma_w^j)} \tag{5.30}$$

$$A^i = \Sigma_{\dot{x}_c w}^i (\Sigma_w^i)^{-1}. \tag{5.31}$$

where $h^i(w) \in [0, 1]$ is a scalar function of w and A^i is a constant square matrix. With (5.30) and (5.31), (5.29) can be simplified as

$$\dot{x}_c = \sum_{i=1}^N h^i(w) \left[\mu_{\dot{x}_c}^i + A^i(w - \mu_w^i) \right] \tag{5.32}$$

Equation (5.32) has the following physical interpretation: the state-varying admittance controller consists of N linear dampers, where each damper has a unique admittance A^i and a preload velocity $\mu_{\dot{x}_c}^i - A^i \mu_w^i$. Equation (5.31) encodes the relation between the admittance A^i and the covariance of Gaussian distribution. The non-linear weight $h^i(w)$ denotes the contribution of each damper to the whole block. If combining multiple dampers into a single non-linear one, then we can define

$$\bar{A} = \sum_{i=1}^N h^i(w) A^i \tag{5.33}$$

$$\bar{\mu}_{\dot{x}} = \sum_{i=1}^N h^i(w)(\mu_{\dot{x}_c}^i - A^i \mu_w^i) \tag{5.34}$$

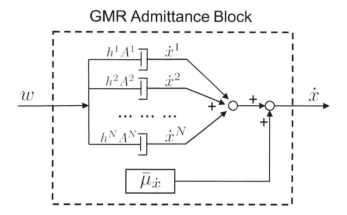

Figure 5.4: Interpretation of GMR from mechanics point of view. The admittance block formulated by GMR consists of N linear dampers with admittance A^i respectively. Each damper has a non-linear weight $h^i(w)$ denoting its contribution to the whole block. The output of the admittance block is generated by the summation of the N dampers' outputs as well as a general preload velocity $\bar{\mu}_{\dot{x}}$.

and $\dot{x}_c$ can be described as

$$\dot{x}_c = \bar{A}w + \bar{\mu}_{\dot{x}} \tag{5.35}$$

where $\bar{A}$ is the general admittance of the block, and $\bar{\mu}_{\dot{x}}$ is the general preload corrective velocity.

To conclude, the structure of GMR has an inherent similarity with physical admittance systems. It can be interpreted as a combination of multiple dampers with different admittances and non-linear weights (see Fig. 5.4). This explicit structure also provides the convenience to analyze the closed-loop stability of the control system.

5.3.3 STABILITY CONDITION OF THE CLOSED-LOOP SYSTEM

For industrial applications, it is critical to guarantee the system stability. This subsection will analyze the stability conditions of the closed loop system with the GMR admittance controller based on Lyapunov theorem [197].

Theorem 5.1

Consider a classic robot force control system shown in Fig. 5.5, which consists of a robot, an admittance control block $\dot{x}_c = \sum_{i=1}^{N} h^i(w_e) \left[\mu_{\dot{x}_c}^i + A^i(w_e - \mu_w^i) \right]$, a velocity

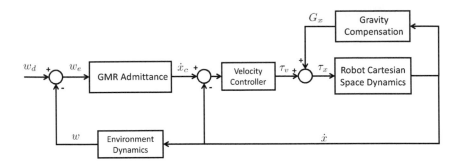

Figure 5.5: Force control block diagram with GMR admittance module.

controller $\tau_v = K_P(\dot{x}_c - \dot{x}) - K_I x$, a gravity compensator, and damped environment $w = K_d \dot{x}$. The closed-loop system is asymptotically stable[1] at $(x, \dot{x}) = 0$ if:

$$K_P \succeq 0 \tag{5.36}$$

$$K_I \succ 0 \tag{5.37}$$

$$\bar{\mu}_{\dot{x}} = 0 \tag{5.38}$$

$$K_P A^i K_d \succeq 0, \qquad \forall i = 1, 2, \cdots, N \tag{5.39}$$

where $\succ 0$ means positive definite, and $\succeq 0$ means positive semi-definite.   ■

Proof. First define a Lyapunov function

$$V = \frac{1}{2} x^\mathsf{T} K_I x + \frac{1}{2} \dot{x}^\mathsf{T} M_x(x) \dot{x} \tag{5.40}$$

where x and $\dot{x}$ are the position and velocity of the robot end-effector. $M_x(x) \succ 0$ is the robot inertia matrix in Cartesian space. By (5.37), $K_I \succ 0$. Therefore $V > 0$. Take time derivative of (5.40),

$$\dot{V} = \dot{x}^\mathsf{T} K_I x + \frac{1}{2} \dot{x}^\mathsf{T} \dot{M}_x(x, \dot{x}) \dot{x} + \dot{x}^\mathsf{T} M_x(x) \ddot{x} \tag{5.41}$$

In Cartesian space, the robot dynamic equation satisfies

$$M_x(x) \ddot{x} + C_x(x, \dot{x}) \dot{x} + G_x(x) = \tau_x \tag{5.42}$$

---

[1] For an autonomous non-linear dynamic system $\dot{x}(t) = f(x(t))$, it is asymptomatically stable at equilibrium point x_e, if it is (1) Lyapunov stable at x_e, and (2) there exists $\delta > 0$ such that if $\|x(0) - x_e\| < \delta$, then $\lim_{t \to \infty} \|x(t) - x_e\| = 0$ [197].

where $C_x(x,\dot{x})$ is the Coriolis matrix, $G_x(x)$ is the gravity term, and τ_x is the control input. It is well-known in robotics that the matrix $\dot{M}_x(x,\dot{x}) - 2C_x(x,\dot{x})$ is skew symmetric[2]

$$\dot{x}^\mathsf{T} \left[\dot{M}_x(x,\dot{x}) - 2C_x(x,\dot{x}) \right] \dot{x} = 0, \quad \forall \dot{x} \tag{5.43}$$

Substitute (5.42) and (5.43) into (5.41), then

$$\dot{V} = \dot{x}^\mathsf{T} K_I x + \frac{1}{2} \dot{x}^\mathsf{T} \left[\dot{M}_x(x,\dot{x}) - 2C_x(x,\dot{x}) \right] \dot{x} + \dot{x}^\mathsf{T} \left[\tau_x - G_x(x) \right]$$
$$= \dot{x}^\mathsf{T} \left[\tau_x - G_x(x) + K_I x \right] \tag{5.44}$$

For the Cartesian space control law,

$$\begin{aligned}
\tau_x &= \tau_v + G_x(x) \\
&= K_P[\dot{x}_c - \dot{x}] - K_I x + G_x(x) \\
&= K_P \{ \sum_{i=1}^{N} h^i(w_e) \left[\mu^i_{\dot{x}_c} + A^i(w_e - \mu^i_w) \right] - \dot{x} \} - K_I x + G_x(x) \\
&= K_P \left[\sum_{i=1}^{N} h^i(w_e) A^i w_e + \bar{\mu}_{\dot{x}} - \dot{x} \right] - K_I x + G_x(x)
\end{aligned} \tag{5.45}$$

The desired contact wrench w_d is zero, therefore $w_e = w_d - K_d \dot{x} = -K_d \dot{x}$. Also from the stability condition, $\bar{\mu}_{\dot{x}} = 0$. Substituting the relation into (5.45), we obtain

$$\tau_x = -K_P \left[\sum_{i=1}^{N} h^i(w_e) A^i K_d + I \right] \dot{x} - K_I x + G_x(x) \tag{5.46}$$

Substituting (5.46) into (5.44), we obtain

$$\dot{V} = -\dot{x}^\mathsf{T} \left[\sum_{i=1}^{N} h^i(w_e) K_P A^i K_d + K_P \right] \dot{x} \tag{5.47}$$

Since $K_P \succeq 0$, $K_P A^i K_d \succeq 0$, and $h^i(w_e) \in [0,1]$, the superposition

$$\sum_{i=1}^{N} h^i(w_e) K_P A^i K_d + K_P$$

should also be positive semi-definite. Therefore $\dot{V} \leq 0$, and the closed-loop system is Lyapunov stable at $(x,\dot{x}) = 0$. By further applying LaSalle's invariance principle [106] on this autonomous system, it is found that the largest invariant set contains the only equilibrium point $(x,\dot{x}) = 0$. Finally, we can conclude that the closed-loop system is asymptotically stable at $(x,\dot{x}) = 0$. $\qquad \square$

---

[2]The skew symmetricity can be proved by Lagrangian mechanics.

Note that the above is a sufficient condition for closed-loop stability, and the assumptions in the theorem are conservative, which require the environment to contain only damping terms. But in many cases, the environment also contains stiffness terms. Besides, the positive semi-definite condition in (5.39) is not easy to achieve. However, if we have the value of K_d, then we can choose $K_P = K_d^{\mathsf{T}}$ to satisfy (5.39), which can be proved by Cholesky decomposition [87].

5.4 CONCLUSION

Imitation is an important capability for robots. On the one hand, it helps the robot to make predictions of the surrounding environment. On the other hand, it enables the robot to mimic others' behaviors in order to accomplish complex tasks. This chapter discusses an approach to make individualized predictions by taking advantage of on-line learning, and a model-free approach for skill learning with varying admittance.

6 Dexterity: Analogy Learning to Expand Robot Skill Sets

6.1 OVERVIEW

Model-based learning and model-free learning are both powerful approaches for teaching robots novel skills. Generally, they follow the same pipeline as shown in Fig. 2.9. During the training stage, the desired control policy is either constructed by modeling or regressed from training data, which tries to capture the system's causality and represent how the system progresses. The achieved policy is supposed to be universal and applicable. Therefore, it is applied during the test stage, i.e., new actions will be generated according to the test states and the trained control policy. This pipeline has been inherently or implicitly utilized in most robot learning methods and achieved great success.

However, for some complicated tasks, neither sophisticated models nor a large amount of training data can be easily obtained for training the control policy. Under this background, this chapter will propose a novel pipeline for robot learning, named *analogy learning*, which is both model-free and data-efficient. This new pipeline focuses on finding correlation, instead of deriving causality. It focuses on finding a transformation between the training and test scenarios, instead of formulating a control policy to relate the training states and training actions. The concept of analogy learning will be introduced in detail in section 6.2. In section 6.3, analogy learning is compared with traditional learning approaches to illustrate its advantages. A robust non-rigid registration method, structure preserved registration (SPR) is introduced in section 6.4 and section 6.5 for implementing analogy learning. Section 6.6 provides summaries and discussions.

6.2 CONCEPT OF ANALOGY LEARNING

Analogy is a cognitive process of transferring information from a particular subject (the training scenario) to another (the test scenario). It is a powerful mechanism for exploiting past experience in planning and problem solving [42].

In general, there are two major stages in a complete analogy pipeline (Fig. 2.10). The first is the reminding stage, where the past scenarios are reminded and compared with the current problem, and the one bearing strong similarity will be identified. The second is the transferring stage, where a transformation process is deployed to retrieve actions that are appropriate for the previous similar scenarios to solve the current problem. Some adaptation is required to update the transformed action to meet the new demands for the current situation.

Analogy is a critical inference method in human cognition. Some robotic researches have implicitly utilized this idea for specific manipulation tasks, such as rope knotting [190]. Unfortunately, analogy learning as a methodology is still underexplored, and has not been broadly applied for learning various skills for robots.

Take robotic grasping as an example. The objective is to find a proper pose of the robot gripper to firmly grasp a specific object. There are already numerous methods developed for robotic grasping. Some methods developed grasping models based on quantitative analysis, such as force closure [161, 173] and form closure [143, 144]. Some methods follow a model-free approach assuming that a grasp policy can be learned from thousand or even million times of trial and error [112, 172, 150]. However, when human beings grasp objects, neither we need a complicated mathematical model for analysis, nor we require million times of learning iterations. There must be some mechanisms involved that enable humans to grasp objects robustly and intuitively with only limited training samples provided. Analogy learning might be one of the hidden mechanisms. For example, when grasping a cup, humans might figure out that the handle is a good location for a robust grasp. In future scenarios where various kinds of cups are given, humans can transfer this experience and regard the handles as good candidates for grasping. There is no physical model involved in the procedure, and only one training sample is required.

Another example is path planning. The robot needs to plan a path connecting the start configuration towards the target configuration without colliding with obstacles. There are model-based approaches (optimization [177, 188], sampling [105, 93]) and model-free approaches (neural network [226, 229], reinforcement learning [99, 98]) for constructing a path planner. However, when human beings are planning paths, we do not need to initialize with an unfeasible path and then use resampling or gradient-descent to gradually avoid collision. Neither do we have to collect much training data and run into obstacles thousands of times before achieving a satisfying policy. Analogy learning might play an important role for humans to plan paths efficiently and intuitively, without sophisticated models or large numbers of trails. As Fig. 6.1 shows, humans can transfer our previous successful experience across scenarios so as to plan new paths. In Figs. 6(a) and (b), the obstacle shapes are different. Instead of regarding the two scenarios independently and solving separately, we can find a spatial transformation from scenario A to scenario B, and transfer the path in A to immediately get a new feasible path in B. During this procedure, no sophisticated model is involved besides a simple spatial transformation. Moreover, only one training sample (scenario A) is required, and can be transferred to various scenarios which contain a single closed obstacle.

6.3 ADVANTAGES OF ANALOGY LEARNING

Analogy learning is an intuitive methodology which human beings utilize inherently in our daily lives. Besides the intuitiveness, applying this approach to robots has

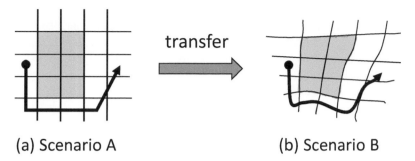

(a) Scenario A   (b) Scenario B

Figure 6.1: Path planning with analogy learning. Grey blocks are obstacles, and blue lines are planned paths. The obstacle shapes are different in two scenarios. A spatial transformation function between the two obstacles can be found and applied to transfer the path for A to get a feasible path for B.

many other benefits including efficiency in learning, robustness to noise and guaranteed collision-free property.

6.3.1   LEARN FROM SMALL DATA

Because of the difficulty in modelling, many methods recently focus on data-driven approaches to regress a control policy for robots from data measurement, such as end-to-end learning [113]. These methods usually require a tremendous amount of data for training before achieving a satisfying result at test. A new phrase, learning from Big Data, has been created for describing this new trend [145]. However, accompanied with the rise of learning from Big Data, it is still unknown why so much data is required for robotic learning. On the one hand, data is not cheap to collect for physical robots, especially industrial manipulators. One may claim to use simulators to virtually simulate a robot. It is true that in a simulation platform, the virtual time can be accelerated arbitrarily to collect more data within the same period. Trial-and-error in a simulator is also acceptable since no physical damage will occur. Unfortunately, creating a sufficiently accurate simulator for physical robots is still a challenging and unsolved problem. As small errors due to the under-modelling accumulate, the simulated robot can quickly diverge from the real-world system. Therefore, reliable data for robot learning still comes from experiments instead of simulations at the current stage. As a result, the data collection process is very expensive. On the other hand, learning from big data is not the straightforward mechanism that human beings are utilizing. Just the opposite, humans learn from small data: only a few samples are enough for humans to learn a skill and to generalize to many other similar situations. It is noticeable that many state-of-the-art methods do succeed in reconstructing the human brain's structure, such as deep neural networks, but one of the human brain's most remarkable features, learning from small data, is regrettably lost.

Why are the current learning methods data hungry? Why can human beings learn from limited data? How can we make robots learn from small data? These fundamental questions have not been answered in the robotics society. In this section, we will try to tackle these problems from the perspective of analogy learning.

In the author's opinion, finding a control policy is like trying to figure out a causality which is universal and generalizable. Causality explains the nature of a system and relates the system states with proper actions. If the causality itself is simple, we can easily construct a model to describe it. If the causality is complex and beyond our ability of modelling, we use model-free approaches to approximate it by high dimensional functions. Since it is a high dimensional function, this approximation inevitably requires big data. If we insist on finding a clear causality using the current high dimensional methodology, we will finally drop to the dilemma of requiring big data for training.

However, it is not always necessary to require a clear causality to solve a problem. Humans can grasp objects much better than robots without understanding force-closure. Humans can plan paths very robustly without understanding space sampling or optimization. Many times, we finish tasks without understanding the physical principles behind. We cannot explain the causality why we take this specific action. Instead, we just intuitively transfer the previous successful experience and apply to the current scenarios, i.e., analogy learning. The core concept of analogy learning is to find the correlation between past scenarios and current situations, instead of the causality between states and actions. It is transferring knowledge, instead of explaining knowledge itself. Its objective is to finish a task by exploiting the past problem-solving actions, instead of explaining how the system works in a fundamental way.

Many state-of-the-art learning methods try to find the causality, of which the procedures are usually data hungry. Our proposed analogy learning approach learns from small data, since it does not seek for the causality. In an extreme case, only one training sample is enough for knowledge transfer. Chapter 9 will show the implementations to teach robots complex skills with limited amount of training samples.

6.3.2   ROBUSTNESS TO INPUT NOISE

For traditional learning methods, the control policy is learned during the training stage and applied to the test stage. However, many of them do not have an inherent functionality to distinguish the input relevance. Taking path planning as an example. As shown in Fig. 6.2, a collision-free path planner is trained by learning methods with the training dataset. At the test stage, the test scenario might not be related to the training scenarios at all, but the trained control policy will still generate a corresponding output and execute it to the system, which is meaningless and might be dangerous to the robot operation.

As a contrast, the analogy learning approach is robust to this input noise. As shown in Fig. 2.12, the reminding stage in analogy learning will first check the similarity between the current input and the training inputs. The following transferring operation will not be executed unless the scenario similarity is large enough. This

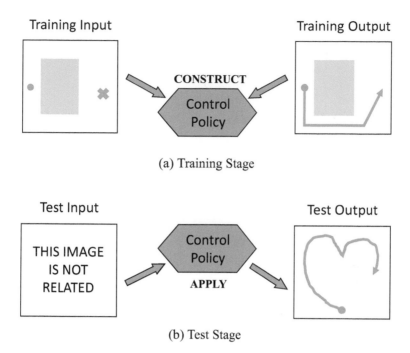

(a) Training Stage

(b) Test Stage

Figure 6.2: Many learning methods do not distinguish the input relevance. (a) At the training stage, a collision-free path planner is trained by learning methods with the training dataset. (b) At the test stage, though the test input is not related to the problem at all, the trained control policy cannot distinguish this noise and will still generate a corresponding output to the system.

self-awareness of what can be handled and what cannot brings additional robustness to the system.

6.3.3 GUARANTEED COLLISION-FREE PROPERTY

Because of the black-box nature in many model-free approaches, the stability and robustness of the system is non-trivial to analyze, which introduces potential risks on the system's safety. As shown in Fig. 6.3, the trained path planner might work well for training data, but could fail and generate infeasible paths for test scenarios.

In contrast, the proposed analogy learning approach can provide a guaranteed collision-free property for robotic motion planning tasks. As illustrated in Fig. 6.1, during the path transformation procedure, if the original path is collision-free, it can be proved that the transformed path is also collision-free. This property is desired especially in those safety-critical scenarios, such as autonomous driving and industrial manipulation. The mathematical proof is provided as follows.

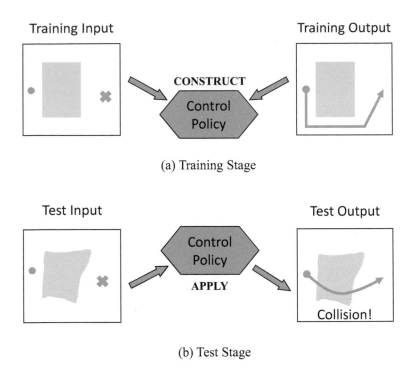

(a) Training Stage

(b) Test Stage

Figure 6.3: Many learned path planners do not guarantee collision-free property. (a) At the training stage, a path planner is trained by learning methods with the training dataset. (b) At the test stage, a new path is generated given new obstacles. However, collision avoidance is not guaranteed.

Theorem 6.1

Given an obstacle (closed set $O \in \mathbb{R}^n$), a collision-free path (closed set $P \in \mathbb{R}^n$), and a homeomorphism (bicontinuous) mapping function ($f : \mathbb{R}^n \to \mathbb{R}^n$), the transformed path $f(P)$ must be collision-free with respect to the transformed obstacle $f(O)$. ∎

Proof. Since the source obstacle and source path are collision-free, we have $O \cap P = \varnothing$. Assume that the transformed path collides with the transformed obstacle, i.e., $f(O) \cap f(P) \neq \varnothing$, then there exists an intersection point $x \in f(P) \cap f(O) \neq \varnothing$. The inverse transform of the intersection point is $x' = f^{-1}(x) \neq \varnothing$. However, $x' \in P \cap O = \varnothing$. Contradiction. Therefore, we must have $f(O) \cap f(P) = \varnothing$, i.e., the transformed path does not collide with the transformed obstacle. □

End of proof.

6.4 STRUCTURE PRESERVED REGISTRATION FOR ANALOGY LEARNING

To implement analogy learning, we need to find a mapping between scenarios, such that: (1) the similarity between the current scenario and the training scenarios can be quantitatively calculated; (2) a transformation function can be constructed so that actions in the past scenario can be transferred.

To satisfy these requirements, *structure preserved registration* (SPR), a non-rigid registration method to register two point sets non-rigidly, is developed in this dissertation. The general idea is to regard the robot training and test scenarios as two point sets. The point correspondence between two sets will be calculated based on a Gaussian mixture probability model. A smooth transformation function will also be formulated to register the corresponding points. This SPR method is shown to be more effective and robust than other state-of-the-art non-rigid registration methods such as TPS-RPM [47], CPD [159] and GLTP [67]. SPR will be utilized in the following chapters to implement the analogy learning approach on robotic grasping, tracking, manipulation and planning.

6.4.1 NODE REGISTRATION WITH GAUSSIAN MIXTURE MODEL

Assume there are two sets of point clouds, source point set $X = \{x_1, x_2, \ldots, x_N\} \in \mathbb{R}^{N \times D}$ and target point set $Y = \{y_1, y_2, \ldots, y_M\} \in \mathbb{R}^{M \times D}$ to represent two environment. N and M are the point numbers in X and Y respectively. D is the point dimension. The objective of SPR is to find a smooth transformation function $\mathcal{T} : \mathbb{R}^D \to \mathbb{R}^D$ to map X to a new position $\bar{X} = \{\bar{x}_1, \bar{x}_2, \ldots, \bar{x}_N\} \in \mathbb{R}^{N \times D}$, such that $\bar{X}$ is well aligned with Y.

Figure 6.4 provides an example of SPR registration. The blue source point set is registered towards the red target point set by a smooth transformation. The point set might contain outliers or exclude partial points.

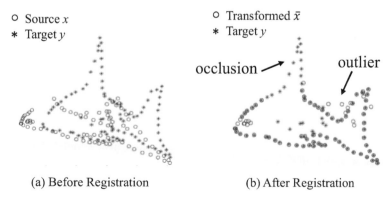

(a) Before Registration   (b) After Registration

Figure 6.4: Illustration of SPR registration. The source point set (blue circle) is registered towards the target point set (red start) by a smooth transformation.

For measuring the alignment level between transformed source set $\bar{X}$ and target set Y, a Gaussian mixture model is constructed, where the transformed points in $\bar{X}$ are regarded as the centroids of multiple Gaussians, and points in Y are random samples from the Gaussian mixtures.

Assume that each Gaussian has equal membership probability $\frac{1}{N}$ and consistent isotropic covariance $\sigma^2 \mathbf{I}$. The probability of point y_m sampled from the Gaussian mixtures can be calculated by:

$$
\begin{aligned}
p(y_m) &= \sum_{n=1}^{N} \frac{1}{N} \mathcal{N}(y_m; \bar{x}_n, \sigma^2 \mathbf{I}) \\
&= \sum_{n=1}^{N} \frac{1}{N} \frac{1}{(2\pi\sigma^2)^{D/2}} \exp(-\frac{\|y_m - \bar{x}_n\|^2}{2\sigma^2})
\end{aligned}
\tag{6.1}
$$

In practice, however, the point set might be noisy and contain outliers. An additional uniform distribution is therefore added to the mixture model to account for noise and outliers. The complete mixture model takes the form:

$$
p(y_m) = \sum_{n=1}^{N+1} p(n)p(y_m|n)
\tag{6.2}
$$

with

$$
p(n) = \begin{cases} (1-\mu)\frac{1}{N}, & n = 1, \ldots, N \\ \mu, & n = N+1 \end{cases}
\tag{6.3}
$$

$$
p(y_m|n) = \begin{cases} \mathcal{N}(y_m; \bar{x}_n, \sigma^2 \mathbf{I}), & n = 1, \ldots, N \\ \frac{1}{M}, & n = N+1 \end{cases}
\tag{6.4}
$$

where μ denotes the weight of the uniform distribution.

$(\bar{x}_n, \sigma^2)$ parameterizes the probability distribution and can be estimated by maximizing the log-likelihood $\mathcal{L}$ of the observation:

$$
\begin{aligned}
\mathcal{L}(\bar{x}_n, \sigma^2|Y) &= \log \prod_{m=1}^{M} p(y_m) \\
&= \sum_{m=1}^{M} \log \left(\sum_{n=1}^{N+1} p(n)p(y_m|n) \right)
\end{aligned}
\tag{6.5}
$$

$$
(\bar{x}_n^*, \sigma^{2*}) = \arg\max_{\bar{x}_n, \sigma^2} \mathcal{L}(\bar{x}_n, \sigma^2|Y)
\tag{6.6}
$$

However, it is non-trivial to solve for $(\bar{x}_n, \sigma^2)$ by directly optimizing over the log-likelihood function $\mathcal{L}$, since there is a summation inside $\log(\cdot)$ which makes convex optimization unfeasible. A complete log-likelihood function Q is therefore constructed,

$$
Q(\bar{x}_n, \sigma^2) = \sum_{m=1}^{M} \sum_{n=1}^{N+1} p(n|y_m) \log (p(n)p(y_m|n))
\tag{6.7}
$$

It can be proved by Jensen's inequality [103] that function Q is the lower bound of function $\mathcal{L}$. Therefore, increasing the value of Q will necessarily increase the value of $\mathcal{L}$ unless it is already at the local optimum. Comparing the structure of Q to that of $\mathcal{L}$, the inside summation is moved to the front of $\log(\cdot)$, which provides a lot of convenience for the following computation.

With the definition of complete log-likelihood function, the EM algorithm [53] which runs expectation step (E-step) and maximization step (M-step) can be utilized to iteratively estimate $(\bar{x}_n, \sigma^2)$ by maximizing Q:

E-step

The expectation step calculates the posteriori probability distribution $p(n|y_m)$ with "old" $(\bar{x}_n, \sigma^2)$ from the last M-step:

$$p(n|y_m) = \frac{\exp(-\frac{\|y_m - \bar{x}_n\|^2}{2\sigma^2})}{\sum_{n=1}^{N} \exp(-\frac{\|y_m - \bar{x}_n\|^2}{2\sigma^2}) + \frac{(2\pi\sigma^2)^{D/2}\mu N}{(1-\mu)M}} \qquad (6.8)$$

M-step

The maximization step plugs $p(n|y_m)$ into the complete log-likelihood function. Ignoring the constants independent of $(\bar{x}_n, \sigma^2)$, we get

$$Q = -\sum_{m=1}^{M}\sum_{n=1}^{N} p(n|y_m)\frac{\|y_m - \bar{x}_n\|^2}{2\sigma^2} - \frac{N_p D}{2}\log(\sigma^2) \qquad (6.9)$$

where $N_p = \sum_{m=1}^{M}\sum_{n=1}^{N} p(n|y_m)$.

Take partial derivative of Q and make $\frac{\partial Q}{\partial \bar{x}_n} = 0$, $\frac{\partial Q}{\partial \sigma^2} = 0$, then a new estimate of $(\bar{x}_n, \sigma^2)$ is achieved:

$$\bar{x}_n = \frac{\sum_{m=1}^{M} p(n|y_m)y_m}{\sum_{m=1}^{M} p(n|y_m)} \qquad (6.10)$$

$$\sigma^2 = \frac{\sum_{m=1}^{M}\sum_{n=1}^{N} p(n|y_m)\|y_m - \bar{x}_n\|^2}{\sum_{m=1}^{M}\sum_{n=1}^{N} p(n|y_m)D} \qquad (6.11)$$

The expectation step and maximization step will be taken alternately until the value of Q converges.

6.4.2 REGULARIZATION ON LOCAL STRUCTURE

Under the aforementioned registration framework, there are no inner constraints between the Gaussian centroids. Each of the centroids will be inevitably registered to the point-rich area to pursue a higher likelihood value (Fig. 6.5b). With respect to a physical object, in contrast, there should exist an inherent topological structure that organizes all the nodes and constrains their motions in sequences.

For this reason, we introduced topological regularization on the transformation, which considers both local and global topology maintenance during transferring X to $\bar{X}$.

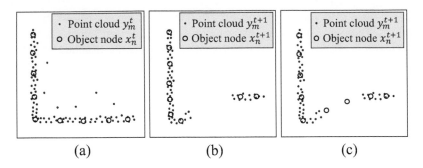

(a)   (b)   (c)

Figure 6.5: Comparison of GMM and SPR on point registration. Blue dot is the point cloud, and red circle is the estimated node position. (1) At time step t, GMM can align X towards Y regardless of noise and outliers. (2) At time step $t+1$, occlusion happens. GMM fails to register points in occlusion area, and the registration on other areas is also seriously disturbed. (3) SPR still works under occlusion because the topological structure is preserved.

Figure 6.6 shows the point x_n and its neighboring points. The local structure around x_n can be characterized as the following weighted sum:

$$x_n = \sum_{i \in I_n} S_{ni} \cdot x_i \tag{6.12}$$

where I_n is the index of the K nearest points to x_n, which can be found efficiently by K-nearest neighbor (KNN, [171]) algorithm. Weight S_{ni} captures the local structure between x_n and its surrounding point x_i. When the point set X is deformed to the new set $\bar{X}$, the absolute point positions are changed. Their inner local structure S_{ni}, however, is expected to be maintained, i.e.:

$$\bar{x}_n \approx \sum_{i \in I_n} S_{ni} \cdot \bar{x}_i \tag{6.13}$$

The optimal combination weights S_{ni} can be achieved by solving the following constrained least square problem:

$$\min_{S_{ni}} \quad \left\| x_n - \sum_{i \in I_n} S_{ni} \cdot x_i \right\|^2$$

$$s.t. \quad \sum_{i \in I_n} S_{ni} = 1 \tag{6.14}$$

Define a difference matrix $D_n = [\cdots, x_i - x_n, \cdots]_{i \in I_n} \in \mathbb{R}^{D \times K}$, and denote the orthogonal projection of one vector $\mathbf{1}_K \in \mathbb{R}^K$ on the null space of D_n as y_0. The optimal solution S_n^* (vectorization of S_{ni}^*) of (6.14) is:

$$S_n^* = \begin{cases} \dfrac{y_0}{\mathbf{1}_K^\mathsf{T} y_0} & , \text{if } y_0 \neq 0 \\[2ex] \dfrac{(D_n^\mathsf{T} D_n)^\dagger \mathbf{1}_K}{\mathbf{1}_K^\mathsf{T} (D_n^\mathsf{T} D_n)^\dagger \mathbf{1}_K} & , \text{else} \end{cases} \tag{6.15}$$

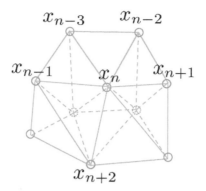

Figure 6.6: Local topology captured by surrounding points.

Unfortunately, the problem of solving S_n^* is not stable if $D_n^\mathsf{T} D_n$ is singular or nearly singular, i.e., a small perturbation on the values of x_i might result in large changes in the solution of S_n^*. Therefore, the local structure is not captured well. From [233], if D_n has L small singular values, there are L suboptimal weight vectors that are linearly independent to each other, each of which partially describes the local structure around point x_n. We can integrate all the L suboptimal weights to characterize the local topology more completely. Denote

$$S_n^{(l)} = (1 - \alpha)S_n^* + VH(:,l), \quad l = 1, \cdots, L \tag{6.16}$$

where V is the right singular vector matrix of D_n corresponding to the L smallest singular values, and $\alpha = \frac{1}{\sqrt{L}}\|V^\mathsf{T} \mathbf{1}_K\|$. H is a Householder matrix given by $H = I - 2hh^\mathsf{T}$ with $h \in \mathbb{R}^L$ defined as:

$$h = \begin{cases} \frac{\alpha \mathbf{1}_L - V^\mathsf{T} \mathbf{1}_K}{\|\alpha \mathbf{1}_L - V^\mathsf{T} \mathbf{1}_K\|} & , \text{if } \alpha \mathbf{1}_L - V^\mathsf{T} \mathbf{1}_K \neq \mathbf{0} \\ 0 & , \text{else} \end{cases} \tag{6.17}$$

It can be proved that $\|D_n S_n^{(l)}\| \leq \|D_n S_n^*\| + \sigma_{K-L+1}(D_n)$ [233]. With these L suboptimal weights, a regularization term for maintaining local topology can be designed as

$$E_{\text{Local}} = \sum_{n=1}^{N} \sum_{l=1}^{L} \| \sum_{i=1}^{N} S_{ni}^{(l)} \bar{x}_i \|^2 \tag{6.18}$$

with $S_{ni}^{(l)} = -1$ if $i = n$, and $S_{ni}^{(l)} = 0$ if $i \notin I_n$. Note that the weight $S_{ni}^{(l)}$ is calculated in X, but the regularization is applied in $\bar{X}$. It is desired that E_{Local} is as small as possible, which indicates the local structure from X is maintained in $\bar{X}$.

6.4.3 REGULARIZATION ON GLOBAL TOPOLOGY

Besides the local structure which regularizes the relative motions between neighbor points, the global topology which regularizes the displacements of the source point set from X to $\bar{X}$ is also critical and needs consideration.

From the macroscopic view, $\bar{X}$ can be assumed to be generated by X by a coherent movement:

$$\bar{x}_n = \mathcal{T}(x_n) \tag{6.19}$$

where $\mathcal{T} : \mathbb{R}^D \to \mathbb{R}^D$ generates globally rigid transformation while also allows locally non-rigid deformation. We would like to find a $\mathcal{T}$ as smooth as possible so as to transform X to $\bar{X}$ coherently. According to the regularization theory [69], the function smoothness can be quantitatively measured by norm $\int_{\mathbb{R}^D} \frac{|T(s)|^2}{G(s)} ds$, where $T(s)$ is the Fourier transform of $\mathcal{T}$. $G(s)$ is symmetric and real with $G(s) \to 0$ as $s \to \infty$. This Fourier-domain norm basically passes $\mathcal{T}$ by a high-pass filter, then measures its remaining power at high frequency. Intuitively, the larger the norm, the more 'oscillation' $\mathcal{T}$ will exhibit, i.e., less smoothness.

Therefore, the regularization term for maintaining global topology can be defined as

$$E_{\text{Global}} = \int_{\mathbb{R}^D} \frac{|T(s)|^2}{G(s)} ds \tag{6.20}$$

A modified likelihood function $\tilde{Q}$ is achieved by involving the regularizations on both local structure and global topology:

$$\tilde{Q} = Q(\bar{x}_n, \sigma^2) - \frac{\tau}{2} E_{\text{Local}} - \frac{\lambda}{2} E_{\text{Global}}$$

$$= Q\left(\mathcal{T}(x_n), \sigma^2\right) - \frac{\tau}{2} \sum_{n=1}^{N} \sum_{l=1}^{L} \| \sum_{i=1}^{N} S_{ni}^{(l)} \mathcal{T}(x_i) \|^2$$

$$- \frac{\lambda}{2} \int_{\mathbb{R}^D} \frac{|T(s)|^2}{G(s)} ds \tag{6.21}$$

Compared to (6.9), the new likelihood function is parameterized by $(\mathcal{T}, \sigma^2)$, instead of $(\bar{x}_n, \sigma^2)$. $\tau \in \mathbb{R}^+$ and $\lambda \in \mathbb{R}^+$ are trade-off weights which balance the data fitting accuracy (from $\bar{X}$ to Y), the local structure maintenance and the global transformation smoothness (from X to $\bar{X}$). The negative signs before τ and λ indicate that a similar local structure and a smoother global transformation are preferred.

6.4.4 CLOSED-FORM SOLUTION FOR TRANSFORMATION FUNCTION

The optimal transformation $\mathcal{T}$ can be found by maximizing the modified likelihood function $\tilde{Q}$. In order to implement analogy learning efficiently, it is critical to perform the aforementioned structure preserved registration as fast as possible. This subsection will prove the existence of the closed-form solution for SPR, which enables registration running in a high frequency.

Theorem 6.2

The maximizer $\mathcal{T}^*$ of the modified likelihood function (6.21) has the following Radial Basis Function (RGF) form:

$$\mathcal{T}(z) = \sum_{n=1}^{N} w_n g(z - x_n) \tag{6.22}$$

where kernel $g(\cdot)$ is the inverse Fourier transform of $G(s)$, and $w_n \in \mathbb{R}^D$ is the kernel weight. ∎

Proof. The transformation function $\mathcal{T}(\cdot)$ can be represented by its inverse Fourier transform:

$$\mathcal{T}(x) = \int_{\mathbb{R}^D} T(s) e^{2\pi i <x,s>} ds \tag{6.23}$$

Substitute (6.23) to (6.21), and take derivative of $\tilde{Q}$ over $T(s)$:

$$
\begin{aligned}
\frac{\delta \tilde{Q}}{\delta T(s)} = {} & \sum_{m=1}^{M}\sum_{n=1}^{N} P_{nm}\frac{1}{\sigma^2}[y_m - \mathcal{T}(x_n)]\frac{\delta \mathcal{T}(x_n)}{\delta T(s)} \\
& - \frac{\lambda}{2}\int_{\mathbb{R}^D}\frac{\delta}{\delta T(s)}\frac{|T(s)|^2}{G(s)} \\
& - \frac{\tau}{2}\sum_{n=1}^{N}\sum_{l=1}^{L}\frac{\delta}{\delta T(s)}\|\sum_{i=1}^{N}S_{ni}^{(l)}\mathcal{T}(x_i)\|^2 \\
= {} & \sum_{m=1}^{M}\sum_{n=1}^{N} P_{nm}\frac{1}{\sigma^2}[y_m - \mathcal{T}(x_n)]e^{2\pi i <x_n,s>} \\
& - \lambda\frac{T(-s)}{G(s)} \\
& - \tau\sum_{n=1}^{N}\sum_{l=1}^{L}\sum_{i=1}^{N}\sum_{j=1}^{N} S_{ni}^{(l)}S_{nj}^{(l)}\mathcal{T}(x_i)e^{2\pi i <x_j,s>}
\end{aligned}
\tag{6.24}
$$

Set $\frac{\delta \tilde{Q}}{\delta T(s)} = 0$ and define two variables:

$$a_n = \sum_{m=1}^{M} P_{nm}\frac{1}{\sigma^2}[y_m - \mathcal{T}(x_n)] \tag{6.25}$$

$$b_n = -\tau \sum_{l=1}^{L}\sum_{j=1}^{N}\sum_{i=1}^{N} S_{ji}^{(l)}S_{jn}^{(l)}\mathcal{T}(x_i) \tag{6.26}$$

The optimal $T^*(s)$ can be formulated as

$$T^*(s) = \tilde{g}(-s) \sum_{n=1}^{N} \frac{a_n + b_n}{\lambda} e^{-2\pi i <x_n, s>} \tag{6.27}$$

Take inverse Fourier transform of $T^*(s)$ to get $\mathcal{T}^*(x)$:

$$\mathcal{T}^*(x) = g(x) * \sum_{n=1}^{N} \frac{a_n + b_n}{\lambda} \delta(x - x_n)$$

$$= \sum_{n=1}^{N} w_n g(x - x_n) \tag{6.28}$$

□

End of proof.

In general, kernel $g(\cdot)$ can take any formulation as long as it is symmetric, positive definite, and $G(s)$ behaves like a low-pass filter. For simplicity, a Gaussian kernel $g(\cdot)$ is chosen, with $g(z - x_n) = \exp(-\frac{\|z - x_n\|^2}{2\beta^2})$. $\beta \in \mathbb{R}^+$ is a manually tuned parameter which controls the rigidity of function $\mathcal{T}$, where large β corresponds to rigid transformation, while small β produces more local deformation.

Theorem 6.3

The modified likelihood function (6.21) is equivalent to:

$$\tilde{Q} = -\frac{1}{2\sigma^2} \text{Tr}(Y^\mathsf{T} d(P^\mathsf{T} \mathbf{1}) Y) + \frac{1}{\sigma^2} \text{Tr}(\mathbf{W}^\mathsf{T} \mathbf{G} P Y)$$

$$- \frac{1}{2\sigma^2} \text{Tr}(\mathbf{W}^\mathsf{T} \mathbf{G} d(P\mathbf{1}) \mathbf{G} \mathbf{W}) - \frac{N_p D}{2} \log(\sigma^2)$$

$$- \frac{\lambda}{2} \text{Tr}(\mathbf{W}^\mathsf{T} \mathbf{G} \mathbf{W}) - \frac{\tau}{2} \text{Tr}(\mathbf{W}^\mathsf{T} \mathbf{G}^\mathsf{T} \Phi \mathbf{G} \mathbf{W}) \tag{6.29}$$

where $\mathbf{G} \in \mathbb{R}^{N \times N}$ is a symmetric positive Gramian matrix with element $\mathbf{G}_{ij} = g(x_i - x_j^{t-1})$ and $\mathbf{W} = [w_1, \ldots, w_N]^\mathsf{T} \in \mathbb{R}^{N \times D}$ is the vectorization of kernel weights in (6.22). $P \in \mathbb{R}^{N \times M}$ is the correspondence matrix with $P_{nm} = p(n|y_m)$. $\mathbf{1}$ is a column vector with all ones. $\Phi = \sum_{l=1}^{L} [S^{(l)T} S^{(l)}]$. $d(\cdot)$ is the diagonalization operation. ∎

Proof. In the modified likelihood (6.21), there are three major terms: the original likelihood term, the local topology regularization term and the global topology regularization term. We will transform each term to get the formulation in (6.29).

First, the original likelihood term. From (6.9),

$$
Q(\mathcal{T}(x_n), \sigma^2) + \frac{N_p D}{2} \log(\sigma^2)
$$

$$
= -\sum_{m=1}^{M} \sum_{n=1}^{N} p(n|y_m) \frac{\|y_m - \mathcal{T}(x_n)\|^2}{2\sigma^2}
$$

$$
= -\sum_{m=1}^{M} \sum_{n=1}^{N} p(n|y_m) \frac{\|y_m^{\mathsf{T}} - G(n,\cdot)^{\mathsf{T}} W\|^2}{2\sigma^2}
$$

$$
= -\frac{1}{2\sigma^2} \sum_{m=1}^{M} \sum_{n=1}^{N} p(n|y_m)[y_m^{\mathsf{T}} y_m + G(n,\cdot)^{\mathsf{T}} W W^{\mathsf{T}} G(n,\cdot) - 2G(n,\cdot)^{\mathsf{T}} W y_m]
$$

$$
= -\frac{1}{2\sigma^2} \{\mathrm{Tr}[Y^{\mathsf{T}} d(P^{\mathsf{T}} \mathbf{1}) Y] + \mathrm{Tr}[W^{\mathsf{T}} G d(P\mathbf{1}) G W] - 2\mathrm{Tr}(W^{\mathsf{T}} GPY)\} \tag{6.30}
$$

Second, the global topology term:

$$
E_{\text{Global}} = \int_{\mathbb{R}^D} \frac{|T(s)|^2}{G(s)} ds
$$

$$
= \int_{\mathbb{R}^D} \frac{T^{\mathsf{T}}(-s) T(s)}{G(s)} ds
$$

$$
= \int_{\mathbb{R}^D} \frac{1}{G(s)} \cdot [G(-s) \sum_{n=1}^{N} w_n^{\mathsf{T}} e^{+2\pi i <x_n, s>}] \cdot [G(s) \sum_{n=1}^{N} w_n e^{-2\pi i <x_n, s>}] ds
$$

$$
= \int_{\mathbb{R}^D} G(s) \sum_{i=1}^{N} \sum_{j=1}^{N} w_i^{\mathsf{T}} w_j e^{+2\pi i <x_j - x_i, s>} ds
$$

$$
= \sum_{i=1}^{N} \sum_{j=1}^{N} g(x_i - x_j) w_i^{\mathsf{T}} w_j
$$

$$
= \mathrm{Tr} \sum_{i=1}^{N} \sum_{j=1}^{N} w_j \mathbf{G}_{ij} w_i^{\mathsf{T}}
$$

$$
= \mathrm{Tr}(W^{\mathsf{T}} GW) \tag{6.31}
$$

Third, the local topology term:

$$
\begin{aligned}
E_{\text{Local}} &= \sum_{n=1}^{N} \sum_{l=1}^{L} \| \sum_{i=1}^{N} S_{ni}^{(l)} x_i^t \|^2 \\
&= \sum_{n=1}^{N} \sum_{l=1}^{L} \text{Tr}[\sum_{i=1}^{N} S_{ni}^{(l)} x_i^{\text{T}} \sum_{j=1}^{N} S_{nj}^{(l)} x_j] \\
&= \sum_{n=1}^{N} \sum_{l=1}^{L} \text{Tr}[X^{\text{T}} S^{(l)}(n,\cdot)^{\text{T}} S^{(l)}(n,\cdot) X] \\
&= \sum_{l=1}^{L} \text{Tr}[X^{\text{T}} S^{(l)T} S^{(l)} X] \\
&= \text{Tr}\{X^{\text{T}} \sum_{l=1}^{L} [S^{(l)T} S^{(l)}] X\} \\
&= \text{Tr}\{W^{\text{T}} G^{\text{T}} \Phi G W\}
\end{aligned}
\tag{6.32}
$$

with

$$
\Phi = \sum_{l=1}^{L} [S^{(l)T} S^{(l)}]
\tag{6.33}
$$

Combine the above three derivations together, then we can get the formulation as listed in (6.29). $\qquad \square$

End of proof.

With the linear formulation of (6.29), the optimal $\mathbf{W}$ and σ^2 can be calculated by taking $\frac{\partial Q}{\partial \mathbf{W}} = 0$ and $\frac{\partial Q}{\partial \sigma^2} = 0$:

$$
\mathbf{W} = [d(P\mathbf{1})\mathbf{G} + \lambda \sigma^2 \mathbf{I} + \tau \sigma^2 \Phi \mathbf{G}]^{-1} PY
\tag{6.34}
$$

$$
\sigma^2 = \frac{1}{N_p D} \{\text{Tr}(Y^{\text{T}} d(P^{\text{T}} \mathbf{1}) Y) - 2\text{Tr}(\mathbf{W}^{\text{T}} \mathbf{G} PY)
$$
$$
+ \text{Tr}(\mathbf{W}^{\text{T}} \mathbf{G} d(P\mathbf{1}) \mathbf{G} \mathbf{W})\}
\tag{6.35}
$$

EM algorithm can also be performed to estimate the parameters iteratively. In E-Step, the posteriori probability distribution P is calculated using the estimated $(\mathbf{W}, \sigma^2)$ from the last M-step. In M-Step, a new estimate of $(\mathbf{W}, \sigma^2)$ is updated by executing the closed-form solutions (6.34) and (6.35).

After $\tilde{Q}$ is converged, the state estimation at time step t can be updated as:

$$
\bar{X} = \mathcal{T}(X) = \mathbf{G}(X)\mathbf{W}
\tag{6.36}
$$

As shown in Fig. 6.5(c), because the topological structure is successfully preserved, the new transformation with SPR is able to produce a reasonable registration under occlusion.

6.4.5   IMPLEMENTATION ACCELERATION

The above registration process requires multiple matrix operations, which might be slow especially when the point number N increases and the state dimension D augments. To accelerate the registration, both Fast Gaussian Transform (FGT) and low rank approximation will be introduced to reduce the computational complexity.

In the M-step in the EM updates, (6.34) and (6.35) involve the matrix-vector products $P\mathbf{1}, P^\mathsf{T}\mathbf{1}$ and matrix-matrix products $P^\mathsf{T}Y$, where P is a variation of Gaussian affinity matrix. Normal production approach takes $\mathcal{O}(MN)$ operations, while with FGT acceleration, the complexity drops to linear $\mathcal{O}(M+N)$. The basic idea is to expand the Gaussians in terms of truncated Hermit expansion for fast computation of the sum of exponentials. More details of the FGT implementation can be found in [75] and [159].

Another bottleneck in our algorithm is the operation of matrix inversion. In (6.34), the matrix $[d(P\mathbf{1})\mathbf{G} + \lambda\sigma^2\mathbf{I} + \tau\sigma^2\Phi\mathbf{G})]$ with dimension $N \times N$ has to be inverted with complexity $\mathcal{O}(N^3)$. A trick of low rank approximation can be applied here to decrease the complexity. First, calculate the eigenvalues and eigenvectors of $\mathbf{G}$ and approximate it by $\mathbf{G} \approx M\Lambda M^\mathsf{T}$, with $\Lambda \in \mathbb{R}^{\sqrt[3]{N} \times \sqrt[3]{N}}$ the diagonal matrix with first $\sqrt[3]{N}$ largest eigenvalues, and $M \in \mathbb{R}^{N \times \sqrt[3]{N}}$ the corresponding eigenvectors. Using Woodbury identity, we get

$$
(AM\Lambda M^\mathsf{T} + \lambda\sigma^2\mathbf{I})^{-1}
$$
$$
= \frac{1}{\lambda\sigma^2}[\mathbf{I} - AM(\Lambda^{-1} + \frac{1}{\lambda\sigma^2}M^\mathsf{T}AM)^{-1}M^\mathsf{T}] \tag{6.37}
$$

where $A = d(P\mathbf{1}) + \tau\sigma^2\Phi$.

In (6.37), the complexity of matrix inversion drops to be linear $\mathcal{O}(N)$.

6.5   EXPERIMENTAL STUDY

Figure 6.7 shows the iterative procedure of registering the blue scissors to a red one by SPR. To have a quantitative comparison with other registration methods, we test the performance of SPR along with those state-of-the-art algorithms, including TPS-RPM [47], CPD [159], GLTP [67], on the commonly used CHUI dataset [48].

Three benchmark tests are performed, including registration with different levels of deformation, occlusions and outliers. For all the tests, the input point sets were first normalized to zero mean and unit variance before registration. The weight μ for uniform distribution was chosen to be 0.1. For global topology regularization, the regularization weight λ and Gaussian kernel's variance β were set as 3.0 and 2.0 respectively. For local structure regularization, the number of nearest neighbor is set as $K = 5$, and its regularization weight is $\tau = 10$. All the point sets were denormalized after registration. Regarding other methods, the default parameters in their papers are utilized, considering they used the same dataset as ours.

As shown in Fig. 6.8, the source point set (blue dot) is desired to be aligned towards the target point set (red dot). From left to right, the target fish and Chinese characters are twisted with larger deformation levels. The proposed SPR algorithm can generate an appropriate transformation function and well align the source to the target at different deformation levels.

Figure 6.9 shows the registration results under occlusions. Note that the target objects (red dots) are partially occluded on purpose, which results in missing points in the point sets. Because SPR maintains the topology of the object during registration, in the red point missing area, the blue points can still be reasonably registered. In the last column of Fig. 6.9, 50% of red points are removed, which shows the registration robustness under massive occlusions.

Similarly, registration is robust when outlier points exist. As shown in Fig. 6.10, dense outliers exist and contaminate the target point set. In the last column, the number of outliers is larger than that of the objects, whereas an accurate alignment between the point sets can still be found.

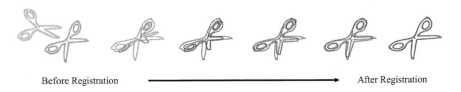

Before Registration ⟶ After Registration

Figure 6.7: Procedure of SPR Registration. The blue scissors are registered towards the red one gradually by SPR.

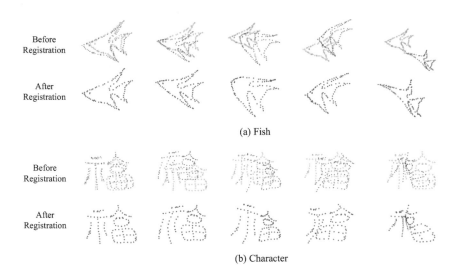

(a) Fish

(b) Character

Figure 6.8: Registration result with different levels of deformations. From left to right, the deformation is increasing.

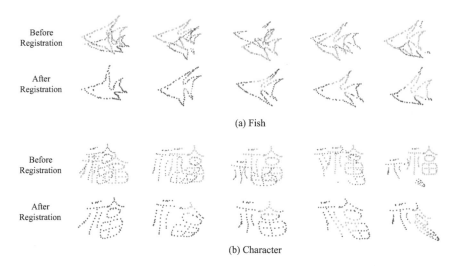

Figure 6.9: Registration result with different levels of occlusions. From left to right, more points are dropped in the source point set.

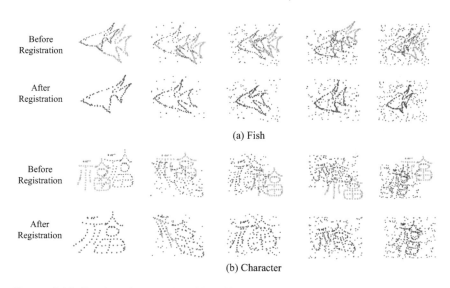

Figure 6.10: Registration result with different levels of outliers. From left to right, more outlier points contaminate the target point set.

Figures 6.11 to 6.13 show the quantitative comparison between SPR and other methods (TPS-RPM, CPD, GLTP) under different levels of deformation and occlusions, respectively. In the CHUI dataset [48], there are 100 tests for each level of deformation, occlusions and outliers. The average registration error (Euclidean dis-

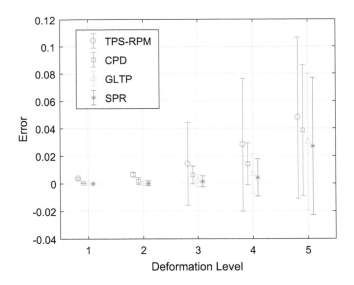

Figure 6.11: Registration error under deformation.

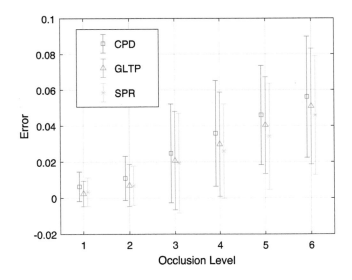

Figure 6.12: Registration error under occlusions.

tance between the transformed source point set and the ground-truth target point set) of the 100 tests is calculated. Note that since TPS-RPM has a large error under occlusions, for the plotting convenience, TPS-RPM is not included in Fig. 6.12. It is shown that SPR has the best result in all the three benchmark tests. The outper-

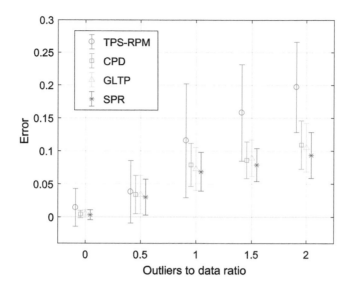

Figure 6.13: Registration error under outliers.

formance is more obvious when there are larger levels of deformation, occlusions and outliers, which suggests that in those circumstances, the regularizations on local structure and global topology play a significant role and provide a more robust registration between point sets.

6.6 CONCLUSION

A new robot learning approach, named analogy learning, is proposed in this chapter. Instead of finding a control policy to relate the system states and corresponding actions, analogy learning tries to find correlation between the current scenario with past training scenarios. A past scenario that bears a strong similarity with the current will be identified (reminding) and the past actions will be transferred to generate a new feasible action (transferring). The benefits of analogy learning, including efficiency in data learning, robustness to input noise and collision-free guarantee are discussed in detail.

To support the reminding and transferring procedures of analogy learning, a novel non-rigid registration method, named structure preserved registration, has been developed. It regards the past and current scenarios as point sets, and finds a smooth transformation function $\mathcal{T}$ to register them to each other. The similarity between the two point sets can be quantitatively analyzed based on the likelihood value Q. The performance of registration is illustrated by a series of experiments to show its robustness under massive levels of deformation, outliers and occlusions.

7 Cooperation: Conflict Resolution during Interactions

7.1 OVERVIEW

In the previous chapters, we discussed methods to design the behavior of a single robot. The performance of the design needs to be evaluated in the context of multi-agent systems. In a multi-agent system, information may be asymmetric in the sense that each agent only knows its own model, cost, and future action, but not others'. In this chapter, we discuss methods to analyze the performance of a multi-agent system under information asymmetry and propose strategies to improve the performance.

In the robotics community, the idea of multi-robot teams has received much attention due to its wide applications in search and rescue, intelligent manufacturing, and unknown environment exploration. Conventionally, distributed control laws are designed for each robot agent [180] by an intelligent designer who has an omniscient perspective. This strategy usually works well on problems with predefined environment and interaction patterns, but has limited extendability, as it is hard to design a universal control law that works in all scenarios, especially when the system topology is time-varying (interactions are happening among different agents at different times).

A more flexible approach is to let the agent's behavior evolve during interactions [131]. The designer only needs to specify a cost and a logic for each agent following the behavior system discussed in section 1.3 and shown in Fig. 1.2a. In this way, the multi-agent system can be self-organized [140]. The cost function determines the behavior of each agent, which can be understood as the agent's character. In certain cases, the cost function of an agent may not be clear to other agents (as a natural situation during HRI). This scenario is called *information asymmetry* [175]. The problem of interest is how to design the logic so that the desired behavior will evolve even in the case of information asymmetry. This is not only important from the control point of view, but also essential in understanding how cooperation evolves among humans and how robots may cooperate with humans.

To address the problem, the closed-loop dynamics of the multi-agent system needs to be analyzed. When agents are reacting to one another under information asymmetry, the multi-agent system may be time-varying, in the sense that the parameters of the system may change throughout time. For example, an agent's cost and logic may be updated by its learning module during the interaction with other agents and the environment. The closed-loop dynamics of a multi-agent system include both *transient response* and *steady-state response*. The transient response describes the system dynamics before the system parameters converge. The steady-state response describes the system dynamics after the system parameters converge, though it is not guaranteed that a multi-agent system will eventually arrive at a steady state. It is worth noting that the definition of the transient and steady-state responses in a multi-agent system are different from their conventional definitions in control theory. In the classic control theory, the transient response and the steady-state response are defined regarding the convergence of the state values, instead of the system parameters. In the steady state of a multi-agent system, the system is no longer time-varying, though the system state may not necessarily converge and is not even guaranteed to converge. In the following discussion, we call the steady state of a multi-agent system an *equilibrium*. In an equilibrium, the system dynamics are time constant, meaning that no agent has incentive to update its control law.

This chapter focuses on analyzing the equilibrium of the multi-agent system. In particular, this chapter tackles the following two questions:

1. whether the system goal (e.g., cooperation among agents) can be achieved in an equilibrium;
2. how to design the agents' behaviors to reach that equilibrium.

There are different notions of equilibria in game theory. When there is no information asymmetry, the steady state response of a multi-agent system is characterized by a Nash equilibrium (NE). In a NE, no agent can be better off by choosing a different control law unilaterally. When there is information asymmetry, the Bayesian Nash equilibrium (BNE) [222] is introduced, which encodes a profile of control laws that minimizes the expected cost for each agent given its belief about others' costs. However, the learning process is not explicitly considered in a BNE. The evolvement of the system (hence the steady state of the system) cannot be characterized by a BNE. To deal with this problem, this chapter introduces a trapped equilibrium (TE) to characterize the steady-state multi-agent system when all learning processes have converged (but may or may not converge to the true values). The aforementioned questions will be answered regarding the TE, e.g., whether the cooperative goal can be achieved in a TE and whether a TE is reachable. In the ideal case, the system should converge to a TE that corresponds to a NE. However, not all NEs are TEs, since some of the NEs are not reachable or stable. And obviously, not all TEs are NEs. For example, a deadlock situation may be a TE, but not a NE. A deadlock is a situation that the system is being static, i.e., $\dot{x} \equiv 0$, while the system goal has not been achieved.

Multiple factors can break cooperation among agents. Theoretically, when agents fail to cooperate, either the system goes to an undesired TE, or the system does not converge to any TE. The failure modes differ under different interaction modes discussed in section 1.2. For systems that require synchronization, e.g., collaborative load transfer, there is usually a unique NE. Cooperation may fail when agents have wrong estimation of others and the system converges to an undesired TE that does not align with the NE, which will be discussed in section 7.3. For systems that require asynchronized actions, e.g., intersection crossing for autonomous vehicles, there are usually multiple NEs. Cooperation may fail when the system is trapped into deadlock (i.e., reach to an undesired TE) or cannot reach consensus on which NE to commit to (i.e., does not settle in steady state), which will be discussed in section 7.4. The strategy to restore cooperation among agents is called *conflict resolution*. Sections 7.3 and 7.4 discuss conflict resolution strategies in the two cases.

The remainder of this chapter is organized as follows. Section 7.2 discusses the closed-loop dynamics of multi-agent systems. Section 7.3 analyzes the effect of information asymmetry on multi-agent systems that require synchronization and discusses strategies to ensure that the system converges to a desired TE. Section 7.4 focuses on intersection crossing for autonomous vehicles and proposes a strategy for conflict resolution using active communication.

7.2 DYNAMICS OF MULTI-AGENT SYSTEMS

This section sets up the mathematical problem for multi-agent interactions following the notations introduced in Chapter 2. In particular, we consider simultaneous games where all agents make their decisions based on a same set of available data, as discussed in section 2.1.2.

7.2.1 SIMULTANEOUS DYNAMIC GAME

Denote the system state as $x \in \mathbb{R}^n$ where n is the dimension of the state space. The agents in the system are indexed as $i = 1, ..., N$, and they provide inputs $u_1, u_2, ..., u_N$ to the system, where $u_i \in \mathbb{R}^{r_i}$ for all i. $U \in \mathbb{R}^r$ is the stack of all u_i's, where $r = \sum_i r_i$. Assume the system dynamics is affine with respect to u_i's, e.g.,

$$x(k+1) = f(x(k)) + \sum_{i=1}^{N} h_i(x(k))u_i(k) \tag{7.1}$$

Then for any agent i, it can only choose its control u_i based on the observations (the current state and the previous inputs of other agents) and its reference goal $G_i \subset \mathbb{R}^n$ by minimizing a cost function $\mathcal{J}_i(x, u_i, G_i)$,

$$u_i = \arg\min \mathcal{J}_i(x, u_i, G_i) \tag{7.2}$$

Since all agents' inputs can affect the system state x, the solution of the optimization problem in (7.2) is indeed a response curve, i.e.,

$$u_i = g_i(x, G_i, u_{-i}), \forall i \tag{7.3}$$

where u_{-i} denotes the inputs of all other agents except the agent i. The intersection of the response curves among all agents is the NE. The resulting control laws $\{u_i^o\}_i$ as a solution of (7.3) are optimal in the sense that an agent cannot get a lower cost by deviating from this control law unilaterally. Figure 7.1 illustrates the NE in the

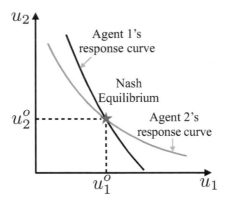

Figure 7.1: The response curve and the Nash Equilibrium.

(u_1, u_2) function space in a two-agent case. However, a NE is only defined for systems with *complete information*, i.e., the cost functions $\mathcal{J}_i$'s or the response curves g_i's are globally known. When those are not globally known, the system is with *incomplete information*. The scenario that the cost functions known to different agents are not identical is called *information asymmetry*. In this chapter, the case that an agent only knows its own cost function is considered, in which case the agents can only act on their estimates of others' response curves.

7.2.2   AGENT STRATEGIES

The reasoning strategy for an agent i consists of the learning strategy (estimating u_{-i}, behavior of other agents) and the control strategy (minimizing its own cost (7.2) based on the estimates). The learning strategy corresponds to the learning module and the control strategy corresponds to the logic module in the behavior system in Fig. 1.2a. From the agent perspective, most learning and control strategies follow a centralized design philosophy. For example, in reinforcement learning [121] and adaptive control [107], an agent's estimation is adjusted if its observation deviates from its prediction. However, in a multi-agent system, where each agent acts on its estimation of the unknowns, the gap between an agent's observation and prediction can be caused by the following two factors:

1. the agent's wrong estimation (wrong prediction);
2. other agent's wrong estimations and subsequent improper actions (deviated observations).

Since those centralized design strategies do not address the second factor, they are likely to produce instability as will be shown in section 7.3 or deadlock as will be shown in section 7.4. To deal with this problem, we can either design a new strategy to compensate the second factor or allow communication among agents.

7.2.3 MULTI-AGENT LEARNING

Under different agent strategies, the multi-agent system can evolve in different ways. We conceptually illustrate the difference between the following two strategies [139]:

1. Blame-Me strategy, which does not compensate for other agents' sub optimality (the second factor discussed above);
2. Blame-All strategy, which compensates for other agents' sub optimality.

Figure 7.2 shows the evolution of the system under the Blame-Me strategy. In Fig. 7.2a, agent i ($i = 1, 2$) chooses its action $\underline{u}_i$ in the intersection of its response curve (the solid curve) and the estimated response curve of the other agent (the dashed curve of the same color). However, the action pair $\underline{u}_1, \underline{u}_2$ in the yellow dot deviates from the agents' predictions of what the other would do. The deviation provides incentives for the agents to update their estimations. From agent 1's perspective, since the deviation is only attributed to its wrong estimation, the updated estimate of agent 2's response curve should go through the observed action pair $\underline{u}_1, \underline{u}_2$ in Fig. 7.2a. A new control law can be chosen based on the new estimation as shown in Fig. 7.2b. Agent 2 goes through the same reasoning. The implicit assumption under this strategy is that other agents are optimal given the observed data, which is a common assumption in data-driven estimation of cost functions [26]. However, this strategy is not ideal as it results in "over-reaction" as the new action pair moves further away from the NE in Fig. 7.2b.[1]

In the Blame-All strategy, it is assumed that other agents are only optimal given their estimates. To predict what others would do, how they are estimating "me" should also be considered. For example, in Fig. 7.3a, the response curve of agent 1 is estimated by both agents (shown as the red and black dashed lines and denoted as Agent 1's ideal response curve). If the two agents follow the same rule of initialization, they will have consensus on the estimates of the response curves. The ideal response curves define a virtual NE (the light blue dot). An agent's best action is to choose the corresponding action on its response curve, assuming others choosing actions in the virtual NE. Since the virtual NE is not the true NE, the observed action pair is still deviating from the predicted. From Agent 1's perspective, the deviation is caused both by its wrong estimation and the Agent 2's improper action. Hence the estimate of Agent 2's response curve is updated only to compensate the gap (between the virtual NE and the action pair) in the u_2 direction, while the gap in the u_1 direction is compensated by updating agent 1's ideal response curve as shown in Fig. 7.3b. In this way, the true response curves are found and the NE is achieved.

---

[1]The Blame-Me strategy actually works for a follower in a sequential game [130]. This is due to the fact that the follower does not need to consider the response of the leader. In a sequential game, only the leader needs to consider the responses of others (i.e., followers). However, in a simultaneous game, all agents need to consider the responses of others.

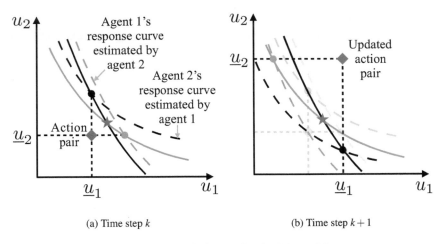

(a) Time step k (b) Time step $k+1$

Figure 7.2: System evolution under the Blame-Me strategy.

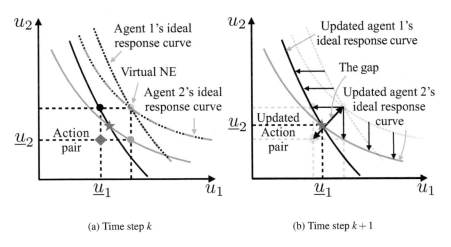

(a) Time step k (b) Time step $k+1$

Figure 7.3: System evolution under the Blame-All strategy.

7.2.4   TRAPPED EQUILIBRIUM

Given different system evolutions discussed above, the problems of interest are: under different strategies, whether the parameter estimates converge to true values and what is the resulting closed-loop system.

A *trapped equilibrium* (TE) is defined as the time-invariant closed loop system where all agents do not have incentives to change their control laws, as their learning processes converge. In Fig. 7.2 and Fig. 7.3, being in the TE means that the yellow dot remains in the same location under the specific reasoning strategy. Under the TE, the closed-loop stability around the goal point will be analyzed to answer the question whether the agents can cooperate.

7.3   COOPERATION UNDER INFORMATION ASYMMETRY

This section discusses strategies to boost cooperation for multi-agent systems that require synchronization under information asymmetry. The scope of analysis is narrowed down to quadratic games, which rise naturally in multi-robot cooperations [193]. In the following discussion, we first formulate the mathematical problem in section 7.3.1. Sections 7.3.2 and 7.3.3 analyze the performance of the Blame-Me strategy and the Blame-All strategy respectively. Section 7.3.4 illustrates the performance on a collaborative load transfer task. The analytical results are directly stated. The proofs can be found in [139].

7.3.1   QUADRATIC GAME AND NASH EQUILIBRIUM

Model and Assumptions

It is assumed that all agents have identical goals, i.e., $G_i = G$ for all i. The system goal G is to drive the system state x to a target state, which is assumed to be at the origin. Consider a quadratic game where every agent i minimizes the following cost

$$\mathcal{J}_i = J_i(x(k)) := x^\mathsf{T}(k+1)Px(k+1) + u_i^\mathsf{T}(k)u_i(k)\theta_i, \qquad (7.4a)$$

$$\text{s.t. } x(k+1) = f(x(k)) + \sum_{i=1}^{N} B_i u_i(k), \qquad (7.4b)$$

where $\theta_i \in \mathbb{R}$ encodes the preference of agent i which is not known to others. The parameter $P \in \mathbb{R}^{n \times n}$ is positive definite, which is chosen such that the closed loop system in the NE is globally asymptotically stable around the origin. The parameter P is identical for all agents and globally known. For simplicity, it is assumed that $B_i := h_i(x(k))$ is constant. Define

$$R := \begin{bmatrix} \theta_1 \mathbf{I}_{r_1} & \cdots & 0 \\ \vdots & \ddots & \vdots \\ 0 & \cdots & \theta_N \mathbf{I}_{r_N} \end{bmatrix} \in \mathbb{R}^{r \times r},$$

$$B := \begin{bmatrix} B_1 & \cdots & B_N \end{bmatrix} \in \mathbb{R}^{n \times r},$$

$$U := \begin{bmatrix} u_1 \\ \vdots \\ u_N \end{bmatrix} \in \mathbb{R}^r.$$

Denote the matrices that contain the diagonal and the off-diagonal entries of $R + B^T PB$ as H_1 and H_2 respectively, where

$$
H_1 \quad := \quad \begin{bmatrix} \theta_1 \mathbf{I}_{r_1} + B_1^T PB_1 & \cdots & 0 \\ \vdots & \ddots & \vdots \\ 0 & \cdots & \theta_N \mathbf{I}_{r_N} + B_N^T PB_N \end{bmatrix} \tag{7.5}
$$

$$
H_2 \quad := \quad \begin{bmatrix} 0 & B_1^T PB_2 & \cdots & B_1^T PB_N \\ B_2^T PB_1 & 0 & \ddots & \vdots \\ \vdots & \ddots & \ddots & B_{N-1}^T PB_N \\ B_N^T PB_1 & \cdots & B_N^T PB_{N-1} & 0 \end{bmatrix} \tag{7.6}
$$

It is assumed that H_1 and H_2 are invertible. If H_1 is singular, then some J_i does not depend on some entries in u_i. If H_2 is singular, then there are no cross coupling terms among some control pairs.

The notations to be used in this section are listed below.

$\underline{x}, \underline{u}_i, \underline{U}$	The observed state and control inputs.
$x^o(k)$	The optimal state at k under complete information given $\underline{x}(k-1)$.
u_i^o, U^o	The optimal control inputs under complete information.
$\delta x, \delta u_i, \delta U$	The deviation of the state and control inputs due to information asymmetry: $\delta x = \underline{x} - x^o$, $\delta u_i = \underline{u}_i - u_i^o$, $\delta U = \underline{U} - U^o$.
$\hat{a}^{(i)}, \tilde{a}^{(i)}$	The estimate and estimation error of a parameter a by agent i such that $\tilde{a}^{(i)} = \hat{a}^{(i)} - a$. For example, $\hat{\theta}_j^{(i)}(k)$ is the estimate of θ_j by agent i at time k and $\tilde{\theta}_j^{(i)}(k) = \hat{\theta}_j^{(i)}(k) - \theta_j$.
$\widehat{a}, \widetilde{a}$	The estimate and estimation error of a parameter a regardless of who makes the estimation.

Benchmark System in the Nash Equilibrium

For any agent i, the response curve $\partial J_i / \partial u_i = 0$ with respect to the cost function (7.4) is

$$
\theta_i u_i(k) = -B_i^T Px(k+1), \tag{7.7}
$$

where others' inputs are implicitly contained in $x(k+1)$. By (7.4b) and using estimates for other agents' inputs, the control input of agent i satisfies

$$
u_i(k) = -[\theta_i I_{r_i} + B_i^T PB_i]^{-1} B_i^T P[f(\underline{x}(k)) + \sum_{j \neq i} B_j \hat{u}_j^{(i)}(k)]. \tag{7.8}
$$

Hence the optimal control law is a linear combination of a state feedback control law

$$
-[\theta_i I_{r_i} + B_i^T PB_i]^{-1} B_i^T P f(\underline{x}(k)),
$$

and a predictive control law

$$-[\theta_i I_{r_i} + B_i^\mathsf{T} PB_i]^{-1} B_i^\mathsf{T} P \sum_{j \neq i} B_j \hat{u}_j^{(i)}(k),$$

where the actions of other agents are predicted. The predictive control law varies under different strategies, to be discussed in the following two sections.

In the NE, agents should have correct predictions, i.e., $\hat{u}_j^{(i)}(k) \equiv u_j(k)$. Stacking the control law (7.8) for all agents and moving the inverse terms to the left, we get

$$H_1 U(k) = -B^\mathsf{T} P f(\underline{x}(k)) - H_2 U(k). \tag{7.9}$$

Hence the saddle solution with respect to (7.4) is

$$U^o(k) = \begin{bmatrix} u_1^o(k) \\ \vdots \\ u_N^o(k) \end{bmatrix} = -\underbrace{[R + B^\mathsf{T} PB]^{-1} B^\mathsf{T} P}_{K} f(\underline{x}(k)), \tag{7.10}$$

where K is the system's feedback gain. The distributed control laws in the NE coincide with the centralized control law w.r.t. the following cost function

$$J(x(k)) = x^\mathsf{T}(k+1) P x(k+1) + U^\mathsf{T}(k) R U(k). \tag{7.11}$$

Denote agent i's feedback gain as

$$K_i = \underbrace{[0, \ldots, 0, \mathbf{I}_{r_i}, 0, \ldots, 0]}_{T_i} K, \tag{7.12}$$

where $T_i \in \mathbb{R}^{r_i \times n}$. Hence the optimal control law for agent i under complete information is

$$u_i^o(k) = -K_i f(\underline{x}(k)). \tag{7.13}$$

Let $y_{ij} \in \mathbb{R}^{r_i \times r_j}$ be the block entries of $[R + B^\mathsf{T} PB]^{-1}$,

$$[R + B^\mathsf{T} PB]^{-1} = \begin{bmatrix} y_{11} & \cdots & y_{1N} \\ \vdots & \ddots & \vdots \\ y_{N1} & \cdots & y_{NN} \end{bmatrix}. \tag{7.14}$$

The parameters y_{ij}'s depend on $\theta_1, \ldots \theta_N$. They encode interactions among agents, since the control strategy of agent i depends on parameters of agent j

$$K_i = \sum_j y_{ij} B_j^\mathsf{T} P, \tag{7.15}$$

and y_{ij} encodes the level of this dependency.

The closed loop system in the NE is

$$\underline{x}(k+1) = [I - BK] f(\underline{x}(k)). \tag{7.16}$$

7.3.2 BLAME-ME STRATEGY

The parameters θ_i's in the cost function (7.4) are not globally known. Nonetheless, the agents should still be able to learn to cooperate with each other. Adaptive algorithms are usually used to let the agent learn the unknown parameters and take actions accordingly. However, if the Blame-Me strategy is adopted, i.e., not compensating the sub optimality of other agents, the closed loop system can be unstable. This section discusses the control and learning algorithms under the Blame-Me strategy and analyzes the properties of the corresponding TEs.

The Adaptive Control Algorithm

In the case that θ_j is only known to agent j, according to the control law (7.10), the adaptive control law of agent i can be written as

$$u_i(k) = -T_i[\hat{R}^{(i)} + B^{\mathsf{T}}PB]^{-1}B^{\mathsf{T}}Pf(\underline{x}(k)), \tag{7.17}$$

where $\hat{R}^{(i)} = \mathrm{diag}(\hat{\theta}_1^{(i)}\mathbf{I}_{m_1}, \dots, \theta_i\mathbf{I}_{m_i}, \dots, \hat{\theta}_N^{(i)}\mathbf{I}_{m_N})$ contains the estimated parameters. Agent i's learning objectives are other agents' response curves (7.7), e.g., the solid curves in Fig. 7.1. Since θ_j is a scalar, it is intuitive to do the parameter estimation using the following equation with reduced order,

$$\theta_j \underline{u}_j^{\mathsf{T}}(k)\underline{u}_j(k) = -\underline{u}_j^{\mathsf{T}}(k)B_j^{\mathsf{T}}P\underline{x}(k+1). \tag{7.18}$$

The parameter θ_j can be learned by agent i using the recursive least square (RLS) [107] discussed in section 5.2.2, i.e.,

$$\hat{\theta}_j^{(i)}(k+1) = \hat{\theta}_j^{(i)}(k) + e_j^{(i)}(k+1)\underline{u}_j^{\mathsf{T}}(k)\underline{u}_j(k)F(k), \tag{7.19}$$
$$e_j^{(i)}(k+1) = -\underline{u}_j^{\mathsf{T}}(k)B_j^{\mathsf{T}}P\underline{x}(k+1) - \hat{\theta}_j^{(i)}(k)\underline{u}_j^{\mathsf{T}}(k)\underline{u}_j(k),$$

where $F(k) \in \mathbb{R}^+$ is the learning gain. For simplicity, $F(k)$ is set to be $(\underline{u}_j^{\mathsf{T}}(k)\underline{u}_j(k))^{-2}$ if $\underline{u}_j(k) \neq 0$ and 0 otherwise. Note that $\underline{u}_j(k)$ is known to agent i at $k+1$.

The Closed Loop System

Proposition 7.1 (Multiplicative uncertainty). *The closed loop dynamics of the multi-agent system when all agents are using the control law (7.17) satisfy*

$$\underline{U}(k) = (\mathbf{I} + \Delta(k))U^o(k), \tag{7.20}$$
$$\underline{x}(k+1) = [\mathbf{I} - B[\mathbf{I} + \Delta(k)]K]f(\underline{x}(k)), \tag{7.21}$$

$$\Delta(k) = -\begin{bmatrix} 0 & \hat{y}_{12}^{(1)}\tilde{\theta}_2^{(1)} & \cdots & \hat{y}_{1N}^{(1)}\tilde{\theta}_N^{(1)} \\ \hat{y}_{21}^{(2)}\tilde{\theta}_1^{(2)} & 0 & \cdots & \hat{y}_{2N}^{(2)}\tilde{\theta}_N^{(2)} \\ \vdots & \vdots & \ddots & \vdots \\ \hat{y}_{N1}^{(N)}\tilde{\theta}_1^{(N)} & \hat{y}_{N2}^{(N)}\tilde{\theta}_2^{(N)} & \cdots & 0 \end{bmatrix}, \tag{7.22}$$

where $\Delta(k) \in \mathbb{R}^{r \times r}$, whose diagonal entries are all zero.

Remark 7.1. *(7.21) shows that the uncertainty introduced by information asymmetry is multiplicative and (7.22) indicates the structure of the uncertainty in the Blame-Me strategy.*

The Trapped Equilibrium

If agent i and agent m use the same learning algorithm with the same initial condition, they should agree on all their estimates of another agent j, e.g., $\hat{\theta}_j^{(i)}(k) = \hat{\theta}_j^{(m)}(k)$, $\forall i, m \neq j$. Define $\widehat{\theta}_j(k) = \hat{\theta}_j^{(i)}(k)$, $\forall i \neq j$. Define $\widehat{R}(k) = \text{diag}(\widehat{\theta}_1(k)\mathbf{I}_{r_1}, ..., \widehat{\theta}_N(k)\mathbf{I}_{r_N})$.

Proposition 7.2 (Convergence of the learning algorithm). *In the system (7.21) with uncertainty (7.22), for all initial conditions, the learning algorithm (7.19) converges, if for some k^*, $\widehat{R}(k^*) = \widehat{R}^e$ and $\Delta(k^*) = \Delta^e$ satisfies*

$$(\Delta^e + \mathbf{I})^{\mathsf{T}}[(\widehat{R}^e + B^{\mathsf{T}}PB)\Delta^e + \widetilde{R}^e] = 0. \tag{7.23}$$

Then for all $k \geq k^*$, $\widehat{R}(k) = \widehat{R}^e$ and $\Delta(k) = \Delta^e$.

Remark 7.2. *(7.22) and (7.23) determine the TEs of the multi-agent system. For example, one TE may be $\Delta^e = \widetilde{R}^e = 0$. Another TE may be $\Delta^e = -[\widehat{R}^e + B^{\mathsf{T}}PB]^{-1}\widetilde{R}^e$. The first TE is efficient as it is identical with the NE. When $\Delta^e \neq 0$, the cooperation under the second TE is inefficient.*

Remark 7.3 (Unstable modes). *When $N = 2$, suppose $B_j^{\mathsf{T}}PB_j = c_j\mathbf{I}$ for some scalar c_j. Then $\widehat{R}^e = -H_1 + R, \Delta^e = H_2^{-1}H_1$ satisfies (7.22) and (7.23), hence defines a TE. Moreover, $(\mathbf{I} + \Delta^e)K = H_2^{-1}(H_1 + H_2)[H_1 + H_2]^{-1}B^{\mathsf{T}}P = H_2^{-1}B^{\mathsf{T}}P$. By (7.21), the closed loop dynamics is*

$$\underline{x}(k+1) = [\mathbf{I} - BH_2^{-1}B^{\mathsf{T}}P]f(\underline{x}(k)), \tag{7.24}$$

which can be unstable when $\|H_2\| \to 0$.

7.3.3 BLAME-ALL STRATEGY

To overcome the instability under the Blame-Me strategy, the Blame-All strategy is proposed to compensate other agents' sub optimality.

The Algorithm

The instability in the Blame-Me strategy is indeed caused by the underlying assumption that agent j is behaving optimally with respect to J_j given the information available to agent i. However, agent j is only optimal given information available to itself.

Putting (7.17) in the form of (7.8), the estimate of $u_j(k)$ in the Blame-Me strategy is equivalent to

$$\hat{u}_j^{(i)}(k) = -[\hat{\theta}_j^{(i)} + B_j^{\mathsf{T}}PB_j]^{-1}[f(\underline{x}(k)) + \sum_{m \neq i,j} B_m\hat{u}_m^{(i)}(k) + B_iu_i(k)], \tag{7.25}$$

which is biased since agent j does not know $u_i(k)$ in advance. The term $u_i(k)$ should also be estimated. And the estimate of $u_j(k)$ should be

$$\hat{u}_j^{(i)}(k) = -[\hat{\theta}_j^{(i)} + B_j^{\mathsf{T}} PB_j]^{-1}[f(\underline{x}(k)) + \sum_{m \neq j} B_m \hat{u}_m^{(i)}(k)], \qquad (7.26)$$

where the estimate $\hat{u}_i^{(i)}(k)$ should be calculated using the same algorithm in estimating other agents' behavior in (7.26). In this case, agent i makes the estimation pretending that it has no prior knowledge of its own parameters.

Following the procedure in obtaining (7.10) from (7.8), (7.26) can be solved for all $j = 1, \cdots, N$,

$$\hat{u}_j^{(i)}(k) = -T_j[\hat{R}^{(i)}(k) + B^{\mathsf{T}} PB]^{-1} B^{\mathsf{T}} Pf(\underline{x}(k)), \qquad (7.27)$$

where $\hat{R}^{(i)} = \mathrm{diag}(\hat{\theta}_1^{(i)} \mathbf{I}_{r_1}, \ldots, \hat{\theta}_i^{(i)} \mathbf{I}_{r_i}, \ldots, \hat{\theta}_N^{(i)} \mathbf{I}_{r_N})$. Specifically, $\hat{\theta}_i^{(i)}$ is the estimate of $\hat{\theta}_i^{(j)}$ for $j \neq i$ made by agent i to compensate for others' wrong estimations of θ_i. When all agents are using the same algorithm with the same initial condition, they should agree on their estimates, i.e., $\hat{R}^{(i)} = \hat{R}^{(j)}$ for all i and j. Then by (7.27), $\hat{u}_j^{(i)}(k)$ does not depend on i. In the following discussion, we omit the superscript in the estimates. The estimate of the parameter matrix is denoted $\hat{R}$. The estimated input is denoted $\hat{u}_j(k)$. Stacking all $\hat{u}_j(k)$, we obtain

$$\hat{U}(k) = -[\hat{R}(k) + B^{\mathsf{T}} PB]^{-1} B^{\mathsf{T}} Pf(\underline{x}(k)). \qquad (7.28)$$

$\hat{U}(k)$ is the virtual NE (the light blue dot in Fig. 7.3) that is consensual among all agents. Plugging (7.28) into (7.8), agent i's control law becomes

$$u_i(k) = -[\theta_i + B_i^{\mathsf{T}} PB_i]^{-1} B_i^{\mathsf{T}} P[f(\underline{x}(k)) + (B - B_i T_i)\hat{U}(k)]. \qquad (7.29)$$

From agent i's perspective, agent j is only optimal given its estimate $\hat{U}(k)$. Equation (7.7) needs to be rewritten as

$$\theta_j u_j(k) = -B_j^{\mathsf{T}} P\hat{x}^{(j)}(k+1), \qquad (7.30)$$

where

$$\hat{x}^{(j)}(k+1) = f(\underline{x}(k)) + (B - B_j T_j)\hat{U}(k) + B_j \underline{u}_j(k). \qquad (7.31)$$

In (7.18) in the Blame-Me strategy, the observed $\underline{x}(k+1)$ is used for the $x(k+1)$ term. However, $\underline{x}(k+1)$ is biased, since it may not be the future state predicted by agent j when it was doing control at time k. In (7.30), on the other hand, the predicted future state $\hat{x}^{(j)}(k+1)$ is used. This calculation is made possible by using the consensus $\hat{U}(k)$. Using the same $F(k)$ in (7.19), the following unbiased RLS algorithm can be formulated,

$$\hat{\theta}_j(k+1) = \hat{\theta}_j(k) + e_j(k+1)\underline{u}_j^{\mathsf{T}}(k)\underline{u}_j(k)F(k), \qquad (7.32)$$
$$e_j(k+1) = -\underline{u}_j^{\mathsf{T}}(k)B_j^{\mathsf{T}} P\left[f(\underline{x}(k)) + (B - B_j T_j)\hat{U}(k) + B_j \underline{u}_j(k)\right] - \hat{\theta}_j(k)\underline{u}_j^{\mathsf{T}}(k)\underline{u}_j(k).$$

The Closed Loop System

Proposition 7.3 (Multiplicative uncertainty). *The closed loop dynamics of the multi-agent system when all agents are using the control law (7.29) follow from (7.20) and (7.21) where*

$$\Delta(k) = H_1^{-1} H_2 [\widehat{R}(k) + B^{\mathsf{T}} P B]^{-1} \widetilde{R}(k). \tag{7.33}$$

The Trapped Equilibrium

Proposition 7.4 (Convergence of the learning algorithm). *In system (7.21) with uncertainty (7.33) and learning algorithm (7.29), the sequence $\{\widehat{\theta}_j(k)\}_k$ is bounded $\forall j = 1,..N$. The sequence converges to θ_j in finite time steps if and only if the set $\{u_j(k) \equiv 0\}$ does not contain any trajectory of the closed loop system.*

Proposition 7.4 implies two TEs under the Blame-All strategy: (1) the parameter estimation converges in finite steps and the system converges to the benchmark system in (7.16); (2) the system is trapped in the set $\cup_j \{u_j(k) \equiv 0\}$. To get rid of the second TE, agent j is allowed to choose a random control input when the optimal input in (7.29) is always zero, which is called a *perturbation* to the system.

Theorem 7.1 (System performance under the Blame-All strategy). *For any N, the closed-loop multi-agent system under the Blame-All strategy will converge to the benchmark system (7.16) in the NE, if the agent j is allowed to perturb the system when the system is trapped in a deadlock set $\{\underline{u}_j(k) \equiv 0, B_j^{\mathsf{T}} P\underline{x}(k) \neq 0\}$.*

Remark 7.4. *The perturbation only works in the Blame-All case, since the convergence is only affected by the initial conditions. In contrast, in the Blame-Me case, the convergence highly depends on the parameter θ_i's, according to Proposition 7.2. The improved performance in the Blame-All case is due to the introduction of the consensus $\widehat{U}(k)$ among all agents, which decouples learning and control strategies.*

7.3.4   EXAMPLE: ROBOT-ROBOT COOPERATION

Consider the case where two mobile robots cooperate in moving a large object as shown in Fig. 7.4a. Suppose there are rigid connections among the robots and the object. Define $x = [p^{\mathsf{T}}, v^{\mathsf{T}}, \alpha, \omega]^{\mathsf{T}} \in \mathbb{R}^6$ where p and v are the position and velocity of the center of mass (which is assumed to be at the middle of the object), α is the orientation of the object and ω is the angular velocity around the center of mass. The inputs of the agents are the forces, e.g., $u_i = F_i \in \mathbb{R}^2$. Then the system dynamics is in the form of (7.1), where m is the mass, M is the inertia, dt is the sampling time, $fr1(\cdot)$ and $fr2(\cdot)$ are the frictions, and l_1 and l_2 are the vectors pointing from the center of mass to the robot 1 and 2 respectively. The task for the robots is to move the object from $(4, 10, 0, 0, -\pi/2, 0)$ to $(0, 0, 0, 0, 0, 0)$ as shown in Fig. 7.4b. The cost function is given by (7.4).

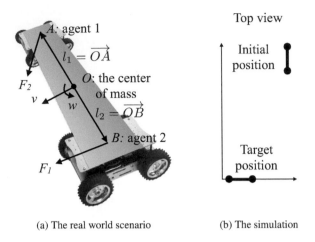

(a) The real world scenario   (b) The simulation

Figure 7.4: Multi-robot cooperation.

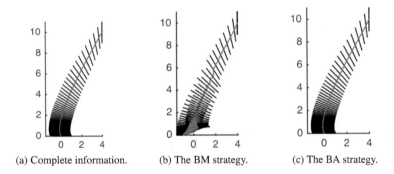

(a) Complete information.   (b) The BM strategy.   (c) The BA strategy.

Figure 7.5: The trajectories under different strategies.

In the simulation, $dt = m = 0.2$, $M = 0.067$, $|l_1| = |l_2| = 1$, $fr1(v) = 0.02v$ and $fr2(\omega) = 0.0067\omega$. The parameters in the cost function are $\theta_1 = 0.3$, $\theta_2 = 0.9$ and The agents initiate the learning process by setting the estimates to be 1. The simulated trajectories are shown in Fig. 7.5 where (a) shows the trajectory under complete information, (b) and (c) show the trajectories under the two strategies when information is asymmetric.

TE under the Blame-Me Strategy: When $\alpha \approx 0$, $B_i = h_i(x)$ is almost constant and $\hat{\theta}_1^{(2)} = \hat{\theta}_2^{(1)} = -2$ is a TE by Remark 7.3. The resulting closed loop matrix $I - BH_2^{-1}B^{\mathsf{T}}P$ in (7.24) has an eigenvalue at 10 (the maximum singular value is 10.8), which causes instability as shown in Fig. 7.5b. Due to the instability, the target position was not reached. Figure 7.6 shows the simulation profile under this strategy and illustrates how the oscillation starts.

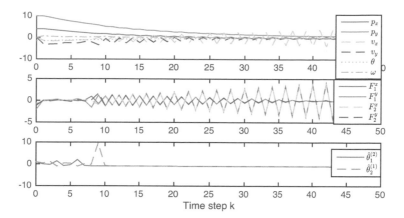

Figure 7.6: The simulation profile under the Blame-Me strategy.

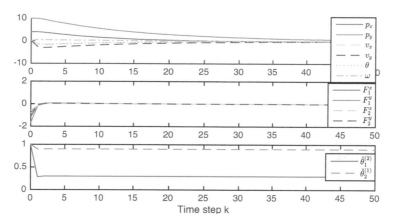

Figure 7.7: The simulation profile under the Blame-All strategy.

TE under the Blame-All Strategy: Figure 7.7 shows the system performance under the Blame-All strategy. The Blame-All strategy outperforms the Blame-Me strategy as the parameter estimation converges in two time steps and the state goes to zero asymptotically, which verifies Theorem 7.1.

7.4 CONFLICT RESOLUTION THROUGH COMMUNICATION

The Blame-All strategy discussed in section 7.3 works for systems that only have one NE. In those problems, cooperation may fail due to information asymmetry.

Without information asymmetry, the future behavior of other agents can be perfectly predicted. Then the unique NE can be achieved.

For systems with multiple NEs, even when there is perfect information on other agents' costs and behavior models, cooperation may still fail if the agents do not have consensus on which NE to commit to. These problems arise naturally in multi-vehicle navigation, or the Chicken game [59]. The game of chicken is a model of conflict for two vehicles in game theory. Each vehicle has two choices: yield or go. If the other vehicle yields, the ego vehicle receives higher payoff to go. If the other vehicle does not yield, the ego vehicle receives higher payoff to yield. The desired case is that one vehicle yields and the other goes. However, as this is a simultaneous game, the vehicles do not know the other vehicle's plan in advance. Hence it is hard for the vehicles to decide whether to yield or to go.

This section discusses a conflict resolution mechanism for systems with multiple NEs, followed by an application of the strategy on multi-vehicle navigation at unmanaged intersections.

7.4.1 CONFLICT RESOLUTION AND NASH EQUILIBRIA

The problem for conflict resolution is discussed below, following the notation in Chapters 2 and 4.

The Agent Perspective

Consider agent i. Its state at time t is $x_i(t)$ and its intention is G_i. An agent has only local view and local information. Let $\mathcal{N}_i$ be a collection of indices of neighboring agents of agent i. At time t_0, given the initial state $x_i(t_0)$, the trajectory $x_i(t)$ for $t > t_0$ needs to be computed. Hence the planning problem for agent i in a T horizon is formulated as,

$$\min_{x_i} J_i(x_i, G_i) := \int_{t_0}^{t_0+T} L_i(x_i(t), G_i)dt + S_i(x_i(t_0+T), G_i), \tag{7.34a}$$

$$\text{s.t. } \dot{x}_i(t) \in \Gamma(x_i(t)), \forall t \in [t_0, t_0+T], \tag{7.34b}$$

$$x_i(t) \in \mathcal{I}(\hat{x}_j^{(i)}(t)), \forall j \in \mathcal{N}_i, \forall t \in [t_0, t_0+T]. \tag{7.34c}$$

Equation (7.34a) is the cost function that evaluates the performance of the trajectory where $L_i(x_i(t), G_i)$ is the run-time cost; and $S_i(x_i(t_0+T), G_i)$ is the terminal cost. Equation (7.34b) ensures that the planned trajectory is dynamically feasible. The set is defined as

$$\Gamma(x_i) := \{\dot{x}_i \mid \exists u_i, \text{s.t.} \dot{x}_i = f(x_i, u_i)\}, \tag{7.35}$$

where f describes the agent dynamics. Equation (7.34c) is the interactive constraint regarding neighboring agents. In the case of collision avoidance, it reduces to a safety constraint that requires the minimum distance to any surrounding vehicle to be greater than a threshold $d_{min} > 0$, i.e.,

$$\mathcal{I}(x_j) := \{x_i \mid d(x_i, x_j) \geq d_{min}\}, \tag{7.36}$$

where function d measures the minimum distance between two agents. $\hat{x}_j^{(i)}(t)$ is the estimate of $x_j(t)$ made by agent i, which is a function of the estimated intention $\hat{G}_j^{(i)}$ of agent j. Both $\hat{G}_j^{(i)}$ and $\hat{x}_j^{(i)}(t)$ can be predicted using methods discussed in section 5.2. Problem (7.34) needs to be solved in every time step when new information is obtained. Methods to solve problem (7.34) are discussed in Chapter 4.

The System Perspective

Suppose there are N agents in the system, indexed from 1 to N. The system state $x(t)$ is a stack of agents' states. All agents in the system solve problem (7.34) for themselves. While all agents would like to stay safe as specified in constraint (7.34c), their goals G_i's may conflict with one another. The system objective is to ensure that the intentions of agents are satisfied efficiently and safely. The optimization problem from the system perspective is,

$$\min_x \sum_{i=1}^{N} w_i J_i(x_i, G_i), \tag{7.37a}$$

$$\text{s.t. } \dot{x}_i(t) \in \Gamma(x_i(t)), \forall i, \forall t \in [t_0, t_0 + T], \tag{7.37b}$$

$$d(x_i(t), x_j(t)) \ge d_{min}, \forall i, j, i \ne j, \forall t \in [t_0, t_0 + T], \tag{7.37c}$$

where the objective function is a weighted sum of the cost function for every agent. The weights w_i's encode priorities among agents. The feasibility constraint (7.37b) is inherited from (7.34b). Equation (7.37c) is the safety constraint for the system. For simplicity, define the safe set $\mathcal{X}_S$ as

$$\mathcal{X}_S := \{x | d(x_i, x_j) \ge d_{min}, \forall i, j, i \ne j\}. \tag{7.38}$$

The system objective provides a measure of efficiency of the agents from the system level. When all agents solve the optimization (7.34) at the same time, it is a simultaneous game. It is ideal that the distributed solutions in (7.34) match the system optima in (7.37). The system optimum is a Nash Equilibrium in the distributed system. However, due to uncertainties in the estimates $\hat{x}_j^{(i)}$ and $\hat{G}_j^{(i)}$ in (7.34c), it is hard to attain the Nash Equilibrium for the distributed system, which may cause inefficiency as will be discussed below.

It is worth noting that the problem in (7.37) is different from the problem in (7.4) in that:

1. the objects of all agents are not identical;
2. the agents are not dynamically decoupled;
3. there are hard constraints in the system.

Inefficiency in the Multi-Agent System

Let x_i^* be the optimal trajectory that minimizes $J(x_i, G_i)$ only considering the dynamic feasibility (7.34b), but not the collision avoidance constraint in (7.34c), i.e.,

$$x_i^* := \arg \min_{\dot{x}_i(t) \in \Gamma(x_i(t))} J_i(x_i, G_i). \tag{7.39}$$

If $[x_1^*(t)^\mathsf{T}, \ldots, x_N^*(t)^\mathsf{T}]^\mathsf{T} \in \mathcal{X}_S$ for all t, there is no conflict among agents. Otherwise, there are conflicts. If there is no conflict, in the ideal case, agents should execute the optimal trajectory x_i^*'s, since individual optima align with system optima in those cases. However, due to uncertainties in the predictions of others, agent i may not dare to execute its optimal trajectory x_i^*. When there are conflicts among agents, the scenario runs into a generalized Chicken game [59]. Executing the optimal trajectory x_i^* can be considered as the choice of "go". Vehicles will collide with each other if none of them "yields", i.e., executing a trajectory different than x_i^*. The situation for $N = 2$ is illustrated in Fig. 7.8. The cost functions and the optimal trajectories x_1^* and x_2^* for agent 1 (V1) and agent 2 (V2) are shown in Fig. 7.8a and Fig. 7.8b respectively. The darker the color, the smaller the cost. The white square is the infeasible set $\mathcal{X}_S^c$ (the complement of $\mathcal{X}_S$), which can be regarded as infinite cost. The pair (x_1^*, x_2^*) does not belong to $\mathcal{X}_S$. Figure 7.8 illustrates the system objective in (7.37a) with $w_1 = w_2$. The system optima are marked as red dots. From the system perspective, either of the agents needs to yield. However, the system optimum is hard to achieve. The system may be trapped in a situation where all vehicles decide to slow down to yield, which is very inefficient. A mechanism is needed to resolve the conflict and break the symmetry.

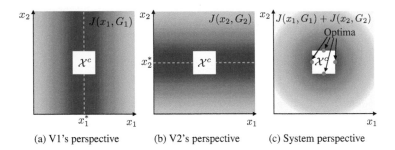

(a) V1's perspective   (b) V2's perspective   (c) System perspective

Figure 7.8: The generalized Chicken game.

7.4.2 COMMUNICATION PROTOCOL AND STRATEGY

While the conflict resolution strategy to be discussed below should work for general situations, we introduce the strategy in light of the problem of multi-vehicle navigation at unmanaged intersections. It is assumed that there is no packet loss or delay in the communication. Moreover, all vehicles are equipped with perfect controllers that can execute the planned trajectories without any error.

Notations

Geometry-wise, we consider general intersections where multiple incoming lanes and multiple outgoing lanes intersect. A conflict zone is formulated when the extensions of two incoming lanes intersect with each other. For example, in the six-leg intersection shown in Fig. 7.9, there are six conflict zones, denoted C_1 to C_6.

We say that vehicle i goes through the conflict zone C_l if there exists $s \in \mathbb{R}^+$ such that $\mathcal{V}_i(x_i^*(s)) \cap C_l \neq \emptyset$ where $\mathcal{V}_i$ denotes the area occupied by vehicle i at state $x_i^*(s)$. Define the segment on path x_i^* that intersects with the conflict zone C_l as $\mathcal{L}_{i,l} := \{s : \mathcal{V}_i(x_i^*(s)) \cap C_l \neq \emptyset\}$. Hence $\mathcal{L}_{i,l} = \emptyset$ if and only if vehicle i does not go through the conflict zone C_l. Denote the set of indices of conflict zones that vehicle i goes through as $\mathcal{A}_i := \{l : \mathcal{L}_{i,l} \neq \emptyset\}$. Then two vehicles i and j go through a same conflict zone if and only if $\mathcal{A}_i \cap \mathcal{A}_j \neq \emptyset$.

Define a discrete state $\mathbf{S}_i$ for vehicle i, where

1. $\mathbf{S}_i = IL$ if vehicle i is in an incoming lane, and is not the first vehicle in the lane;
2. $\mathbf{S}_i = FIL$ if vehicle i is in an incoming lane, and is the first vehicle in the lane;
3. $\mathbf{S}_i = I$ if vehicle i is at the intersection;
4. $\mathbf{S}_i = OL$ if vehicle i is in an outgoing lane.

Vehicle i may enter the control area with $\mathbf{S}_i = IL$ or FIL. $\mathbf{S}_i$ can transit from IL to FIL, from FIL to I, and from I to OL, i.e., becoming the first in lane, entering the intersection and leaving the intersection. It can leave the control area when $\mathbf{S}_i = OL$. For any vehicle i such that $\mathbf{S}_i = IL$ or OL, its front vehicle is denoted $\mathbf{F}_i$.

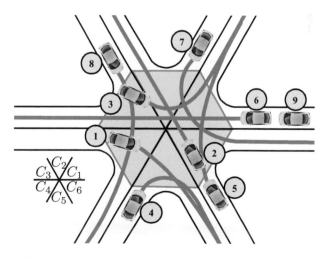

Figure 7.9: Conflict resolution at unmanaged intersections: nine vehicles in a six-way intersection.

Communication Protocol

Communication helps the ego vehicle better determine the constraint (7.34c). Indeed, instead of estimating others' trajectories $\hat{x}_j^{(i)}$, what really matters to the ego vehicle is the time that others occupy the conflict zones. The communication protocol is designed to be that all vehicles should broadcast their estimated time to occupy the conflict zones once they enter a control area of the intersection, along with other basic information such as vehicle ID, current state (position, heading, speed, and $\mathbf{S}_i$) and time stayed in the control area.

Overview of Strategy

Based on the broadcasted information, the vehicles can achieve a consensus on the passing order and compute desired time slots to pass the conflict zones, which are then taken as temporal constraints on the vehicles' trajectories. This naturally breaks the problem into two parts as shown in Fig. 7.10a:

1. decision making – determination of passing order, hence temporal constraints;
2. motion planning – computation of trajectory.

The time flow and the coordination among different modules are shown in Fig. 7.10b. It is assumed that all vehicles are synchronized. At time step $n - 1$, the estimated time interval $[\mathbb{T}_{j,l}^{in,n-1}, \mathbb{T}_{j,l}^{out,n-1}]$ for vehicle j to occupy C_l is broadcasted for all j and l. At time step n, vehicle i evaluates all information received from other vehicles. Then vehicle i computes the desired time slots $[T_{i,l}^{in,n}, T_{i,l}^{out,n}]$ to pass the conflict zone C_l for all l in the decision maker. The desired time slots are then sent to the motion planner as temporal constraints. After motion planning, the planned trajectory is sent to the controller for execution and the estimated time slots to occupy the conflict zones given the new trajectory, i.e., $[\mathbb{T}_{i,l}^{in,n}, \mathbb{T}_{i,l}^{out,n}]$ for all l, are broadcasted to other vehicles. In Fig. 7.10a and Fig. 7.10b, the dashed lines represent information flow. In Fig. 7.10a, detailed architecture is only shown for vehicle i, while other vehicles are assumed to have the same architecture. The thick bars in Fig. 7.10b illustrate the busy time in each module. The decision maker runs sequentially with the motion planner, while the two run in parallel with the controller. The sampling time of the system dt represents one cycle in the decision maker and the motion planner. The controller can have a higher sampling frequency.

The decision maker resolves the conflicts based on the received information. The basic strategy is that whoever arrives first in a conflict zone goes first.[2] However, this strategy may create deadlocks when one vehicle arrives earlier in one conflict zone, while the other vehicle arrives earlier in another conflict zone. A tie breaking mechanism is needed in the decision making module. In this following discussion, we

---

[2]This is different from the strategies discussed in [18], which only considers the arrival time at the intersection.

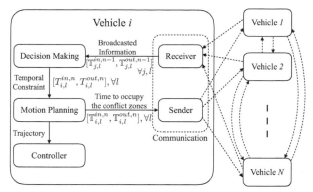

(a) System block diagram. $\mathbb{T}$ represents the broadcasted information. T represents the temporal constraints computed internally. The first subscript is vehicle index. The second subscript is conflict zone index. The second superscript is time step.

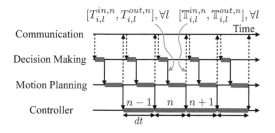

(b) Time flow of execution.

Figure 7.10: Architecture of the conflict resolution mechanism.

introduce a general methodology to deal with distributed scheduling with multiple conflict zones, which is summarized in Algorithm 2.

Note that if $\mathbf{S}_i = IL$, it is physically "constrained" by its front vehicle and should yield all vehicles that its front vehicle yields. The decisions when $\mathbf{S}_i = FIL$ or I are the most important as conflicts usually occur among vehicles in these two states. When $\mathbf{S}_i = OL$, the vehicle no longer needs to compute the desired time interval. However, its information should be broadcasted in order for the proceeding vehicles to follow the lane safely. In the following discussion, we focus on vehicle i with $\mathbf{S}_i = FIL$ or I.

Spatial Conflict

We say that there is a spatial conflict between vehicle i and j if and only if their paths go through a same conflict zone. Consider the scenario shown in Fig. 7.9. By adding links between any pair of vehicles that have spatial conflicts, we formulate an undirected graph as shown in Fig. 7.11a, where every node represents one vehicle.

Algorithm 2 The decision making algorithm for vehicle i to resolve conflicts.

  Input: Broadcasted information $[\mathbb{T}_{j,l}^{in,n-1}, \mathbb{T}_{j,l}^{out,n-1}], \forall j, l$.

  Output: Temporal constraints $[T_{i,l}^{in,n}, T_{i,l}^{out,n}], \forall l$.

Initialize, $n = 0$

while in the control area **do**

    Receive other's information $\mathbb{T}_{j,l}^{in,n-1}, \mathbb{T}_{j,l}^{out,n-1}$

    Initialize $\mathcal{Y}_i = \emptyset$, $T_{i,l}^{in,n} = -\infty$, $T_{i,l}^{out,n} = \infty$

    if $\mathbf{S}_i = IL$ **then**

        i yields its front vehicle ($\mathcal{Y}_i = \{\mathbf{F}_i\}$)

    end if

    for j that has spatial conflicts with i ($j \in \mathcal{U}_i$) **do**

        if j has temporal advantage over i ($j \in \mathcal{V}_i$) **then**

            if $\nexists \text{Tie}(i, j)$ or j has priority over i **then**

                i yields j ($\mathcal{Y}_i = \mathcal{Y}_i \cup \{j\}$)

            end if

        end if

        if i has temporal advantage over j ($i \in \mathcal{V}_j$) **then**

            if $\exists \text{Tie}(j, i)$ and j has priority over i **then**

                i yields j ($\mathcal{Y}_i = \mathcal{Y}_i \cup \{j\}$)

            end if

        end if

    end for

    for j that i yields ($j \in \mathcal{Y}_i$) **do**

        for C_l that both i and j traverse ($l \in \mathcal{A}_i \cap \mathcal{A}_j$) **do**

            $T_{i,l}^{in,n} = \max\{T_{i,l}^{in,n}, \mathbb{T}_{j,l}^{out,n-1} + \Delta_{\mathbf{S}_i}\}$

        end for

    end for

    $n = n + 1$

end while

Whenever there is a link between two vehicles, we need to decide which vehicle goes first. In other words, the undirected graph needs to be transformed into a directed graph as shown in Fig. 7.11c so that the passing order is decided as: a vehicle should yield its successor. Denote the set of vehicles that have spatial conflicts with vehicle i as

$$\mathcal{U}_i := \{j | \mathbf{S}_j = FIL \text{ or } I, \mathcal{A}_i \cap \mathcal{A}_j \neq \emptyset\}. \tag{7.40}$$

The graph in Fig. 7.11a is denoted as $\mathcal{U} := \cup_i \cup_{j \in \mathcal{U}_i} (i, j)$, where (i, j) represents a link between i and j. There is an undirected link between any i and j such that $j \in \mathcal{U}_i$. In literature, this graph is identified as a conflict graph [165]. Finding the optimal passing order regarding the conflict graph is NP-hard. The method presented below is a heuristic approach which finds one feasible passing order in linear time.

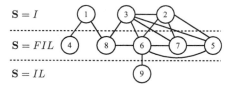

(a) Graph of spatial conflicts $\mathcal{U}$ at one time step. Link $(i, j) \in \mathcal{U}$ implies that vehicle j has spatial conflicts with vehicle i at some conflict zone.

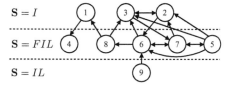

(b) Graph of temporal advantages $\mathcal{V}$ at one time step. Link $(i, j) \in \mathcal{V}$ implies that vehicle j has temporal advantages over vehicle i.

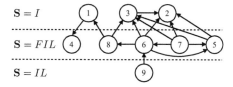

(c) Graph of passing order $\mathcal{Y}$ at one time step. Link $(i, j) \in \mathcal{Y}$ implies that vehicle i yields vehicle j.

Figure 7.11: The conflict graphs.

Temporal Advantage

At time step n, we say that vehicle $j \in \mathcal{U}_i$ has temporal advantage over vehicle i, if one of the following conditions holds,

1. $\mathbf{S}_j = I, \mathbf{S}_i = FIL$ and vehicle i leaves some conflict zones later than vehicle j enters, i.e.,

$$\exists l \in \mathcal{A}_i \cap \mathcal{A}_j, \mathbb{T}_{i,l}^{out,n-1} > \mathbb{T}_{j,l}^{in,n-1}. \tag{7.41}$$

2. $\mathbf{S}_j = FIL, \mathbf{S}_i = I$ and vehicle j leaves all conflict zones earlier than vehicle i enters, i.e.,

$$\forall l \in \mathcal{A}_i \cap \mathcal{A}_j, \mathbb{T}_{j,l}^{out,n-1} \leq \mathbb{T}_{i,l}^{in,n-1}. \tag{7.42}$$

3. $\mathbf{S}_j = \mathbf{S}_i = FIL$ or I and vehicle j enters some conflict zones earlier than vehicle i, i.e.,

$$\exists l \in \mathcal{A}_i \cap \mathcal{A}_j, \mathbb{T}_{j,l}^{in,n-1} \leq \mathbb{T}_{i,l}^{in,n-1}. \tag{7.43}$$

According to the above definitions, for vehicle i and j with different discrete states, either i or j should have temporal advantage over the other. If vehicle i and j have the same discrete state, it is possible that both i and j have temporal advantages over the other. Denote the set of vehicles that have temporal advantages over vehicle i at time step n as $\mathcal{V}_i^n$. The superscript n in the following discussion is ignored for simplicity. Obviously, $\mathcal{V}_i \subset \mathcal{U}_i$. And $\mathcal{V} := \cup_i \cup_{j \in \mathcal{V}_i} (i,j)$ is a directed graph as shown in Fig. 7.11b, where there is a directed link from any i to any $j \in \mathcal{V}_i$. However, there are cycles among nodes with the same discrete state, e.g., between nodes 6 and 7, as well as among nodes with different discrete states, e.g., among nodes 6, 7 and 3. If vehicles yield each other according to the graph, there are deadlocks.

Tie Breaking

For any vehicle i and vehicle $j \in \mathcal{V}_i$, it is called a tie if

1. $\mathbf{S}_i = \mathbf{S}_j$ and there exists a sequence of vehicles $\{q_m\}_1^M$ with $q_1 = i$, $q_M = j$, $M \geq 2$ and $\mathbf{S}_{q_m} = \mathbf{S}_i$ for all m such that $q_m \in \mathcal{V}_{q_{m+1}}$ for $m = 1, \ldots, M-1$.
2. $\mathbf{S}_i = I$, $\mathbf{S}_j = FIL$ and there exists a sequence of vehicles $\{q_m\}_1^M$ with $q_1 = i$, $q_M = j$ and $M \geq 2$ such that $q_m \in \mathcal{V}_{q_{m+1}}$ for $m = 1, \ldots, M-1$.

Let $\text{Tie}(i,j)$ denote all these sequences. The relationship in a tie is neither symmetric nor exclusive, i.e., $\exists \text{Tie}(i,j)$ neither implies $\exists \text{Tie}(j,i)$ nor $\not\exists \text{Tie}(j,i)$. In Fig. 7.11b, there is a tie from node 5 to node 6 via the sequence $\{5,7,6\}$, but not a tie from node 6 to node 5 since $5 \notin \mathcal{V}_6$. There is a tie from node 2 and node 3 via the sequence $\{2,3\}$ and a tie from node 3 and node 2 via the sequence $\{3,2\}$.

We assume that each vehicle has a unique priority score P, which is related to the weights in (7.37a). For example, the priority score of a fire truck is higher than that of a passenger vehicle. We say vehicle i has priority over j if there exists a sequence in $\text{Tie}(i,j)$ such that $P(i) > P(k)$ for all $k \neq i$ in the sequence. Vehicles in the intersection should always have priority over vehicles in the incoming lanes. For vehicles in the same discrete state, the order implied by the priority score should not change over time. If vehicle i has priority over $j \in \mathcal{V}_i$, instead of i yielding j, vehicle j should yield vehicle i, although vehicle j has temporal advantage. For example, in Fig. 7.11, we identify the score P with the vehicle index. Since node 5 has priority in the sequence $\{5,7,6\}$, the link from 5 to 6 is reversed in Fig. 7.11c. Since there is a tie between node 2 and node 6, the link from 2 to 6 is also reversed in Fig. 7.11c.

Passing Sequence

After tie breaking, all those remaining links for vehicle i represent the set of vehicles that vehicle i decides to yield at time step n, which is denoted by $\mathcal{Y}_i^n$. The superscript n in the following discussion is ignored for simplicity. Indeed, $\mathcal{Y} := \cup_i \cup_{j \in \mathcal{Y}_i} (i,j)$ is a directed graph as shown in Fig. 7.11c, which encodes the order for the vehicles to pass the intersection. It is not necessary for vehicle i to construct the whole graphs $\mathcal{U}$ and $\mathcal{V}$ to determine $\mathcal{Y}_i$. For example, vehicle 4 in Fig. 7.11 only needs to compute $\mathcal{U}_4$ and $\mathcal{V}_4$ locally to determine that $\mathcal{Y}_4 = \emptyset$. Those local decisions form the passing sequence globally. In the extreme case, the passing order follows the order specified by the priority score. If all vehicles agree on the above tie breaking mechanism, they can solve the conflicts even if the vehicles plan and control their motions differently.

According to Algorithm 2, if $\mathbf{S}_i = IL$, the vehicle i yields its front vehicle, i.e., $\mathcal{Y}_i = \{\mathbf{F}_i\}$, as shown by vehicle 9 in Fig. 7.11c. If vehicle i decides to yield vehicle j, then for all $l \in \mathcal{A}_i \cap \mathcal{A}_j$, we set

$$T_{i,l}^{in,n} \geq \mathbb{T}_{j,l}^{out,n-1} + \Delta_{\mathbf{S}_i}, \tag{7.44}$$

where $\Delta_{\mathbf{S}_i}$ is a margin to increase the robustness of the algorithm, which is chosen such that $\Delta_{IL} > \Delta_{FIL} > \Delta_I$. Δ_{IL} is chosen to be greater than Δ_{FIL} in order to let the leading vehicles have temporal advantages over vehicles in the middle of other lanes. For example, vehicle 7 will have temporal advantage over vehicle 9 in Fig. 7.11c. Similarly, Δ_{FIL} is chosen to be greater than Δ_I.

Motion Planning

At time step n, given the temporal constraint $[T_{i,l}^{in,n}, T_{i,l}^{out,n}]$ specified by the decision maker, the motion planning problem (7.34) for vehicle i is formulated as a speed profile planning problem,

$$\min_{s_i} J(x_i^*(s_i), G_i) \tag{7.45a}$$

$$\text{s.t. } \dot{x}_i^*(s_i)\dot{s}_i \in \Gamma(x_i^*(s_i)), \tag{7.45b}$$

$$s_i(t) \notin \mathcal{L}_{i,l}, \forall t \notin [T_{i,l}^{in,n}, T_{i,l}^{out,n}], \forall l, \tag{7.45c}$$

where $s_i(t)$ is the speed profile that needs to be optimized. Constraint (7.45c) specifies that the vehicle should only enter the conflict zone C_l in the time interval $[T_{i,l}^{in,n}, T_{i,l}^{out,n}]$. For simplicity, the constraint for car following is omitted in presentation. The problem (7.45) can be solved via temporal optimization as discussed in Chapter 4.

Given the optimal solution s_i^* of (7.45), the expected time slot $[\mathbb{T}_{i,l}^{in,n}, \mathbb{T}_{i,l}^{out,n}]$ for vehicle i to occupy the conflict zone C_l is computed as

$$\mathbb{T}_{i,l}^{in,n} := \min_{s_i^*(t) \in \mathcal{L}_{i,l}} t \geq T_{i,l}^{in,n}, \quad \mathbb{T}_{i,l}^{out,n} := \max_{s_i^*(t) \in \mathcal{L}_{i,l}} t \leq T_{i,l}^{out,n}. \tag{7.46}$$

If $\mathcal{L}_{i,l} = \emptyset$, then $\mathbb{T}_{i,l}^{in,n} := \infty$ and $\mathbb{T}_{i,l}^{out,n} := -\infty$. If vehicle i has entered or left C_l, then $\mathbb{T}_{i,l}^{in,n}$ and $\mathbb{T}_{i,l}^{out,n}$ are chosen as the time that it entered or left C_l respectively.

It is proved in [125] that the conflict resolution strategy solves the conflicts safely and efficiently in real time, in the sense that

1. the passing order is completely determined;
2. there is no deadlock for any pair of vehicles that pass through a same conflict zone at every time step;
3. a stable consensus on conflict-resolution can be reached in finite time steps.

7.4.3 EXAMPLE: UNMANAGED INTERSECTIONS

In this section, we illustrate the performance of the distributed conflict resolution mechanism in traffic simulation. The sampling time in the system is chosen to be $dt = 0.1s$. The robustness margins are chosen as $\Delta_{IL} = 0.5s$, $\Delta_{FIL} = 0.3s$ and $\Delta_I = 0.1s$. The priority score P for a vehicle is chosen to be the time that the vehicle stays in the control area. If there is a tie, then the vehicle with smaller ID has the priority. The cost function of the vehicle penalizes (1) the deviation from a target speed, (2) the magnitude of acceleration or deceleration, (3) the magnitude of jerk, and 4) the time spent in every conflict zone. The target speed varies for different vehicles.

The simulation environment is a four-way intersection as shown in Fig. 7.12a. There is only one incoming lane and one outgoing lane in every direction. Four conflict zones are identified. The control area is the whole graph.

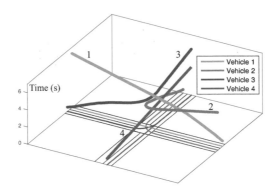

(a) Executed trajectories in the time-augmented space.

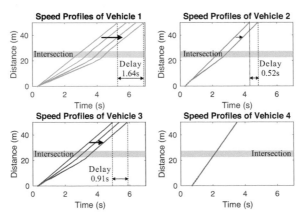

(b) Planned speed profiles in different time steps.

Figure 7.12: Resolved speed profiles and trajectories at the unmanaged intersection.

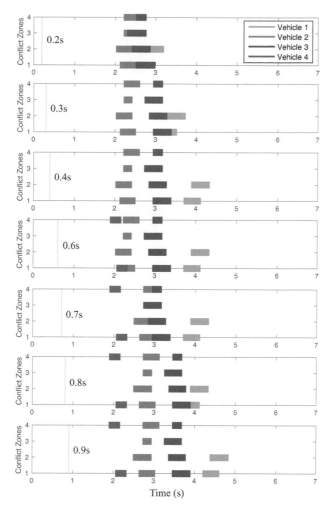

Figure 7.13: Information broadcasting for conflict resolution. The scenario in 0.5 s is omitted since it is the same as the scenario in 0.4 s.

There are four vehicles. The paths and the executed trajectories are shown in the time-augmented space in Fig. 7.12a. The planned speed profiles in different time steps are shown in Fig. 7.12b. The left most speed profile in every subplot is the traffic-free speed profile and the others are the replanned speed profiles given the temporal constraints. Figure 7.13 shows the expected time intervals (the colored thick bars) for the vehicles to occupy the conflict zones. The thin vertical line indicates the current time. Vehicles 1, 2 and 3 enter the control area at the same time. According to the traffic-free speed profiles, there are temporal conflicts between vehicle 1 and vehicle 3 in conflict zones 1 and 2, and between vehicle 2 and vehicle 3 in all conflict zones. Since vehicle 2 has the temporal advantage over vehicle 3, vehicle 3 yields vehicle 2. Similarly, vehicle 1 yields vehicle 3. It takes two time steps for the conflicts to be resolved. At 0.6 s, vehicle 4 enters, which creates new conflicts. The system settles down after 3 time steps as shown in Fig. 7.13. The planned speed profiles change accordingly as shown in Fig. 7.12b. The right most speed profile in each subplot is the executed speed profile.

It is shown in [125] that the method discussed here outperforms other existing distributed intersection protocols due to its flexibility in adjusting the local "yield or go" decisions, hence the passing order in real time.

7.5 CONCLUSION

This chapter investigated methods to resolve conflicts and facilitate cooperation in multi-agent systems. In particular, two kinds of problems were investigated: control and learning strategies in quadratic games with information asymmetry, communication-enabled conflict resolution strategies for systems with multiple NEs.

Part III

Applications

8 Human-Robot Co-existence: Space-Sharing Interactions

8.1 OVERVIEW

For successful human-robot interactions, it is important to ensure that humans and robots co-exist in harmony. Such co-existence is called space-sharing interactions. This chapter discusses two scenarios in space-sharing interactions: co-existence of human workers and robot workers on factory floors, and co-existence of human-driven vehicles and autonomous vehicles on public roads.

In modern factories, human workers and robots are two major workforces. For safety concerns, the two are normally separated with robots confined in metal cages, which limits the productivity as well as the flexibility of production lines. In recent years, attention has been directed to remove the cages so that human workers and robots may collaborate to create a human-robot co-existing factory [43]. The potential benefits of co-robots are huge and extensive, e.g., they may be placed in human-robot teams in flexible production lines [102] as shown in Fig. 8.1, where robot arms and human workers cooperate in handling workpieces, and automated guided vehicles (AGV) coexist with human workers to facilitate factory logistics [218]. Automotive manufacturers Volkswagen and BMW [5] have taken the lead to introduce human-robot cooperation in final assembly lines in 2013. In the factories of the future, more and more interactions among humans and industrial robots are anticipated to take place. In such environments, it is important to ensure safe co-existence among humans and robots [204]. The safety issues have attracted attentions from standardization bodies [82], as well as from major robot manufacturers including Kuka, Fanuc, Nachi, Yaskawa, Adept, and ABB [13]. Several safe cooperative robots or co-robots have been released, such as UR5 from Universal Robots (Denmark) [2] which is implemented by Volkswagen and BMW, Baxter from Rethink Robotics (US) [1], NextAge from Kawada (Japan) [3] and WorkerBot from Pi4˙Robotics GmbH (Germany) [33]. However, most of these researches and products focus on intrinsic safety, i.e., safety in mechanical design [88], actuation [235] and low level motion control [147]. Safety during social interactions with humans, which is key to successful interactions, still needs to be explored.

Automated driving is widely viewed as a promising technology to revolutionize today's transportation systems [37], so as to free the human drivers, ease the road congestion and lower the fuel consumption among other benefits. Substantial re-

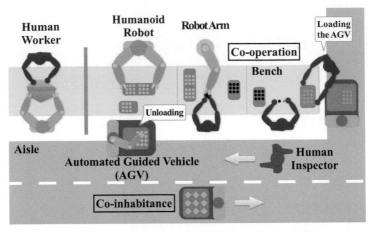

Figure 8.1: Human-robot collaboration and co-inhabitance in future production lines.

search efforts are directed into this field from research groups and companies [7], which are also encouraged by policy makers [184]. When the automated vehicle drives on public roads, safety is a big concern. While existing technologies can assure high fidelity sensing and robust control, the challenges lie in the interactions between the automated vehicle and other road participants such as manually driven vehicles and pedestrians [219]. Conservative strategies such as "braking when collision is imminent", known as the Automatic Emergency Braking (AEB) function in existing models [4], may not be the best actions in many cases. Considering the dynamics and future courses of surrounding vehicles, the automated vehicle has various choices for a safe maneuver. For example, the vehicle may slow down to keep a safe headway until the headway reaches the safe limit; the vehicle may steer to the left or right to avoid a collision; and the vehicle may even speed up if it can get out of a dangerous zone by doing so. For active safety, the vehicle should be able to figure out all safe actions and choose the best action among all choices. In order for these vehicles to interact safely and efficiently with other road participants, the behavior of the automated vehicles should be carefully designed.

The methodologies discussed in the theory part of this book will be applied to these two scenarios, in particular, the safe control strategy discussed in Chapter 3 and the human behavior prediction method discussed in Chapter 5. Section 8.2 discusses the robot safe interaction system (RSIS) to ensure safe co-existence between human workers and robot workers. Section 8.3 discusses the robustly-safe automated driving system (ROAD) to ensure safe interactions with other road participants.

8.2 ROBOT SAFE INTERACTION SYSTEM FOR INDUSTRIAL ROBOTS

In order to make the industrial co-robots human-friendly, they should be equipped with the abilities [81] to:

1. collect environmental data and interpret such data;
2. adapt to different tasks and different environments;
3. tailor themselves to the human workers' needs.

The first ability is a perception problem, while the second and third are control problems that will be studied in this section. The challenges for control are (1) coping with complex and time-varying human motion, and (2) assurance of real time safety without sacrificing efficiency. A constrained optimal control problem is formulated to describe this problem mathematically. And a modularized controller architecture will be discussed as a variation of the architecture discussed in section 2.2.1. The modularized architecture

1. treats the efficiency goal and the safety goal separately and allows more freedom in designing robot behaviors;
2. is compatible with existing robot motion control algorithms and can deal with complicated robot dynamics;
3. guarantees real time safety;
4. is good for parallel computation.

The remainder of the section is organized as follows. Section 8.2.1 describes the constrained optimization problem. Section 8.2.2 discusses the controller architecture to solve the optimization problem, together with the design considerations of each module. Case studies with robot arms are performed in section 8.2.3. Section 8.2.4 concludes the section.

8.2.1   THE OPTIMIZATION PROBLEM

As shown in Fig. 8.1, co-robots can co-operate as well as co-inhabit with human workers. This section addresses safety in contactless space sharing interactions. Since the interaction is contactless, robots and humans are independent from one another in the sense that the humans' inputs will not affect the robots' dynamics in the open loop. Nonetheless, humans and robots are coupled together in the closed loop, since they will react to others' motions.

Problem Formulation

Denote the state of the robot of interest as $x_R \in \mathbb{R}^n$ and the robot's control input as $u_R \in \mathbb{R}^m$ where $n, m \in \mathbb{N}$. Assume the robot dynamics is affine,[1] i.e.,

$$\dot{x}_R = f^x(x_R) + f^u(x_R)u_R. \tag{8.1}$$

---

[1] Any system can have an affine form through dynamic extension. Suppose $\dot{x}_R = f(x_R, u_R)$. Define $x_R^e := [x_R^T, u_R^T]^T$. Let the new control input be $u_R^e = \dot{u}_R$. Then the new system

$$\dot{x}_R^e = \begin{bmatrix} f(x_R^e) \\ 0 \end{bmatrix} + \begin{bmatrix} 0 \\ 1 \end{bmatrix} u_R^e,$$

is affine.

The task or the goal for the robot is denoted as G_R, which can be (1) a settle point in the Cartesian space, e.g., a workpiece the robot needs to get, (2) a settle point in the configuration space, e.g., a posture, (3) a path in the Cartesian space or (4) a trajectory in the configuration space.

The robot should fulfill the aforementioned tasks safely. Let x_H be the state of humans and other moving robots in the system, which are indexed as $H = \{1, 2, \cdots, N\}$. Then the system state is $x = [x_R^\mathsf{T}, x_H^\mathsf{T}]^\mathsf{T}$. The safe set is a subset of the collision free state space, i.e., $\mathcal{X}_S \subset \{x : d(x_H, x_R) > 0\}$ where d measures the minimum distance among the robot, the humans and all other moving robots. Given the human configuration, the constraint on the robot's state space $R_S(x_H)$ is a projection of $\mathcal{X}_S$, i.e., $R_S(x_H) := \{x_R : [x_R^\mathsf{T}, x_H^\mathsf{T}]^\mathsf{T} \in \mathcal{X}_S\}$, which is time varying with x_H. These notations align with the notations in Chapter 3. Two steps are needed to safely control the robot motion:

1. predicting the human motion;
2. finding the safe region for the robot based on the prediction.

The Optimization Problem

The requirement of the co-robot is to finish the tasks G_R efficiently while staying in the safe region $R_S(x_H)$, which leads to the following optimization problem [56]:

$$\min_{u_R} J(x_R, u_R, G_R), \tag{8.2a}$$

$$s.t.\ u_R \in \Omega, x_R \in \Gamma, \dot{x}_R = f^x(x_R) + f^u(x_R)u_R, \tag{8.2b}$$

$$x_R \in R_S(x_H), \tag{8.2c}$$

where J is a goal-related cost function to ensure efficiency; Ω is the constraint on control inputs; Γ is the state space constraint, e.g., joint limits, stationary obstacles. The problem is hard to solve since the safety constraint $R_S(x_H)$ is nonlinear, nonconvex and time varying with unknown dynamics.

The safe control method discussed in Chapter 3 and the human motion prediction method discussed in Chapter 5 will be incorporated to solve the problem. These methods will be generalized and a modularized architecture, called the robot safe interaction system (RSIS), will be discussed in the next section.

8.2.2 THE ROBOT SAFE INTERACTION SYSTEM

As shown in Fig. 8.2, the controller is designed as a parallel combination of a baseline controller and a safety controller. The baseline controller solves (8.2a) to (8.2b), which is time-invariant and can be solved offline. The safety controller enforces the time varying safety constraint (8.2c), which computes whether the baseline control signal is safe to execute or not (in the "Safety Constraint" and the "Criteria" module) based on the predictions made in the human motion predictor, and what the modification signal should be (in the "Control Modification" module). Each module will be elaborated below.

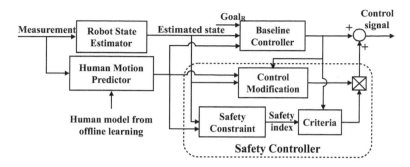

Figure 8.2: The modularized architecture of the robot safe interaction system.

The Baseline Controller

The baseline controller solves (8.2a) to (8.2b), which is similar to the controller in use when the robot is working in the cage. The cost function is usually designed to be quadratic which penalizes the error to the goal and the magnitude of the control input. For example, when G_R is a trajectory, the cost function can be designed to be $J = \int_0^\mathsf{T} [(x_R - G_R)^\mathsf{T} P(x_R - G_R) + u_R^\mathsf{T} R u_R] dt$ where P and R are positive definite matrices. The control policy can be obtained by solving the problem offline. The collision avoidance algorithms discussed in Chapter 4 can be used to avoid stationary obstacles described by the constraint $x_R \in \Gamma$. Depending on the robot's goal, various other algorithms can be incorporated in the baseline controller, i.e., iterative learning control [35]. This controller is included to ensure that the robot can still perform the tasks properly when the safety constraint $R_S(x_H)$ is satisfied.

The Human Model and the Human Motion Predictor

In different applications, the human body should be represented at various levels of details. For AGVs, mobile robots, and planar arms, since the interactions with humans happen in 2D, a human can be tracked as a rigid body in the 2D plane with the state x_H being the position and the velocity of the center of mass and the rotation around it. For robot arms that interact with humans in 3D, the choice of the human model depends on the distance between the human and the robot. When the robot arm and the human are far apart, the human should also be treated as one rigid body to simplify the computation. In the close proximity, however, the human's limb movements should be considered. As shown in Fig. 8.3a, the human is modeled as a connection of ten rigid parts: part 1 is the head; part 2 is the trunk; part 3, 4, 5, and 6 are upper limbs; and part 7, 8, 9, and 10 are lower limbs. The joint positions can be tracked using 3D sensors [192]. The human's state x_H can be described by a combination of the states of all rigid parts.

The prediction of future human motion x_H needs to be done in two steps: inference of the human's goal G_H and prediction of the trajectory to the goal. Once the goal is

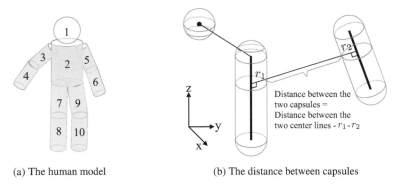

(a) The human model (b) The distance between capsules

Figure 8.3: The human model and the capsules.

identified (using the method of model selection discussed in Chapter 5), a linearized reaction model can be assumed for trajectory prediction [231],

$$\dot{x}_H = A x_H + B_1 G_H + B_2 x_R + w_H, \tag{8.3}$$

where w_H is the noise, A, B_1, and B_2 are unknown matrix parameters which encode the dependence of future human motion on the human's current posture, the human's goal and the robot motion. Those parameters can be identified using parameter identification algorithms discussed in Chapter 5, while the prediction can be made using the identified parameters. Note that to account for human's time varying behaviors, the parameters should be identified online. This method is based on the assumption that human does not "change" very fast. Moreover, to reduce the number of unknown parameters, key features that affect human motion can be identified through offline analysis of human behavior. Those low dimension features $\{f_i\}$ can be used in the model (8.3) to replace the high dimension states x_H and x_R. Then (8.3) becomes $\dot{x}_H = \sum_i A_i f_i + B_1 G_H + w_H$.

The Safety Controller

The Safety Index

The safe set $\mathcal{X}_S$ is a collision free subspace in the system's state space, which depends on the relative distance among humans and robots. Since humans and robots have complicated geometric features, simple geometric representations are needed for efficient online distance calculation. Ellipsoids [216] were used previously. However, it's hard to obtain the distance between two ellipsoids analytically. To reduce the computation load, capsules (or spherocylinders) [149], which consist of a cylinder body and two hemisphere ends, are introduced to bound the geometric figures as shown in Fig. 8.3a. A sphere is considered as a generalized capsule with the length of the cylinder being zero. The distance between two capsules can be calculated analytically, which equals to the distance between their center lines minus their radiuses as shown in Fig. 8.3b. In the case of a sphere, the center line reduces to a point. In this way, the relative distance among complicated geometric objects can be calculated just using several skeletons and points. The skeleton representation is also ideal for tracking the human motion.

Given the capsules, $\mathcal{X}_S$ can be specified by designing the required minimum distances among the capsules. The design should not be too conservative, while larger buffer volumes are needed to bound critical body parts such as the head and the trunk, as shown in Fig. 8.3a. The safe set is designed to be

$$\mathcal{X}_S = \{x : \frac{d(p_{ij}, x_R)}{d_{ij,min}} > 1, \forall i = 1, \cdots, 10, \forall j \in H\}, \tag{8.4}$$

where $d(p_{ij}, x_R)$ measures the minimum distance from the capsule of body part i on the human j to the capsules of the robot R. The constant $d_{ij,min} \in \mathbb{R}^+$ is the designed minimum safe distance. The constant $d_{1j,min}$ captures the distance requirement to the human head, which should be designed large enough.

To describe the safe set $\mathcal{X}_S$ efficiently, a safety index ϕ is introduced, which is a Lyapunov-like function over the system's state space as illustrated in Fig. 3.1b that satisfies the three conditions discussed in section 3.2. The safety index for the safe set in (8.4) is designed as:

$$\phi = 1 + \gamma - (d^*)^c - k_1 \dot{d}^* - \cdots - k_{l-1}(d^*)^{(l-1)}, \tag{8.5}$$

where $d^* = \frac{d(p_{i^* j^*}, x_R)}{d_{i^* j^*, min}}$ and the capsule i^* on the human j^* is the capsule that contains the closest point to the robot R. The closest point is called the critical point. $l \in \mathbb{N}$ is the relative degree from the function $d(\cdot, x_R)$ to u_R in the Lie derivative sense. In most applications, $l = 2$ since the robot's control input can affect joint acceleration. $c > 1$ is a tunable parameter, while larger c implies heavier penalties on small relative distance. $\gamma > 0$ is a safety margin that can uniformly enlarge the capsules in Fig. 8.3a. $k_1, \cdots, k_{l-1}$ are tunable parameters that need to satisfy the condition that all roots of $1 + k_1 s + \cdots + k_{l-1} s^{l-1} = 0$ should be on the negative real axis in the complex plane. The higher order terms of d^* are included to ensure that the robot does not approach the boundary of the safe set at a large velocity, so that the state can always be maintained in the safe set even if there are constraints on the robot control input, i.e., $u_R \in \Omega$.

The Criteria

Given the safety index, the criteria module determines whether or not a modification signal should be added to the baseline controller. There are two kinds of criteria:

1. $\phi(t) \geq 0$, which defines a reactive safety behavior, i.e., the control signal is modified once the safety constraint is violated;
2. $\phi(t + \Delta t) \geq 0$, which defines a forward-looking safety behavior, i.e., the safety controller considers whether the safety constraint will be violated Δt time ahead.

The prediction in the second criterion is made upon the estimated human dynamics and the baseline control law. In the case when the prediction of future x_H has a distribution, the modification signal should be added when the probability for the second criterion to happen is non-trivial, e.g., $\mathbb{P}(\{\phi(t+\Delta t) \geq 0\}) \geq \varepsilon$ for some $\varepsilon \in (0,1)$.

The Set of Safe Control and the Control Modification

The set of safe control U_R^S is the equivalent safety constraint on the control space, i.e., the set of control that can drive the system state into the safe set as shown in Fig. 3.1b. According to (3.5) to (3.8), the set of safe control when $\phi \geq 0$ is

$$U_R^S = \{u_R : \frac{\partial \phi}{\partial x_R} f^u(x_R)u_R \leq -\eta - \frac{\partial \phi}{\partial x_R} f^x(x_R) - \frac{\partial \phi}{\partial x_H}\dot{x}_H\}, \qquad (8.6)$$

where $\eta \in \mathbb{R}^+$ is a margin and $\dot{x}_H$ comes from a human motion predictor. When $\dot{x}_H$ has a distribution, let Γ be the compact set that contains major probability mass of $\dot{x}_H$, e.g., $\mathbb{P}(\{\dot{x}_H \in \Gamma\}) \geq 1 - \varepsilon$ for a small ε. Then the inequality in (8.6) should hold for all $\dot{x}_H \in \Gamma$ as discussed in section 3.4.

The non-convex state space constraint $R_S(x_H)$ is then transferred to a linear constraint on the control space in (8.6). In this way, the modification signal is the optimal value to be added to the baseline control law so that the final control lies in the set of safe control,

$$\Delta u_R = \arg\min_{u_R^o + u \in U_R^S \cap \Omega \cap U_\Gamma} u^\top Q u, \qquad (8.7)$$

where $Q \in \mathbb{R}^{m \times m}$ is positive definite which determines a metric on the robot's control space. To obtain optimality, Q should be close enough to the metric imposed by the cost function J in (8.2a), e.g., $Q \approx d^2 J/du_R^2$ where J is assumed to be convex in u_R. U_Γ is the equivalent constraint on the control space of the state space constraint Γ, which can be constructed following the same procedure of constructing U_R^S. Equation (8.7) is a convex optimization problem and is easy to solve. In the case that $U_R^S \cap \Omega \cap U_\Gamma$ is empty, a smaller margin η can be chosen so that the feasible control set becomes nonempty.

8.2.3 CASE STUDIES

Case studies are performed to evaluate RSIS on scenarios shown in Fig. 8.1. The cases for AGVs and mobile robots are studied in Chapter 3. In this section, the interactions among robot arms and humans will be studied. We use the virtual reality-based human-in-the-loop evaluation platform discussed in section 1.5.2. The results with the dummy-robot evaluation platform can be found in [123].

The Fanuc M16iB robot arm is used in the case study as shown in Fig. 1.1a and the simulation environment is shown in Fig. 8.4a. Capsules are calculated for both the human and the robot as shown in Fig. 8.4b. The radius of the capsules are designed such that one uniform minimum distance requirement $d_{min} = 0.2\,\text{m}$

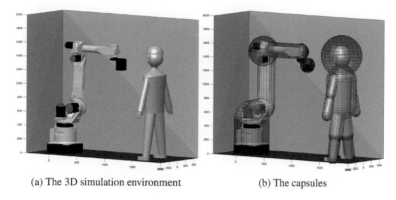

(a) The 3D simulation environment   (b) The capsules

Figure 8.4: The 6DoF robot arm and the simulation environment.

can be used for all capsules. Denote the robot state as $x_R = [\theta^T, \dot{\theta}^T]^T \in \mathbb{R}^{12}$ where $\theta = [\theta_1, \theta_2, \theta_3, \theta_4, \theta_5, \theta_6]^T \in \mathbb{R}^6$ are the joint angles. The control input $u_R = \ddot{\theta} \in \mathbb{R}^6$ is the joint acceleration. The control modification is done in the kinematic level. A perfect low level tracking controller is assumed. The state space equation of the robot arm is linear:

$$\dot{x}_R = A x_R + B u_R, \tag{8.8}$$

where $A_R = \begin{bmatrix} \mathbf{0}_6 & \mathbf{I}_6 \\ \mathbf{0}_6 & \mathbf{0}_6 \end{bmatrix}$ and $B_R = \begin{bmatrix} \mathbf{0}_6 \\ \mathbf{I}_6 \end{bmatrix}$.

The goal G_R is to follow a path in the Cartesian space. The baseline controller is a feedback and feedforward controller. The human is moving around the robot arm. The safety index is designed to be similar to (3.24),

$$\phi = D - d^2 - \dot{d}, \tag{8.9}$$

where the distance function d between 3D capsules is computed analytically [57]. The sampling frequency is 20 Hz. The forward-looking criteria is used. The set of safe control is $U_R^S(k) = \{u_R(k) : \phi(k+1) < 0\}$, and $Q = \mathbf{I}$.

The simulation results are shown in Fig. 8.5 and Fig. 8.6. The first plot in Fig. 8.5 shows the critical capsule ID on the robot arm that contains the closest point to the human and the second plot shows the critical capsule ID on the human that contains the closest point to the robot. During interactions, those critical points changed from time to time. The minimum distance between the human and the robot is shown in the third figure, which was maintained above the danger zone during the simulation. The tracking error is shown in the fourth plot. When the human was far from the robot, perfect tracking was achieved from $k = 100$ to $k = 200$. When the human went close to the robot at $k = 230$, the safety controller took over and moved the robot arm away from the human, at the cost of large tracking error. The snapshots at $k = 230 : 5 : 250$ are shown in Fig. 8.6. In this simulation, the human subject can only control the planar movement of the dummy.

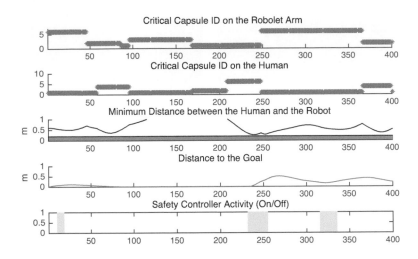

Figure 8.5: The simulation profile of the 6DoF robot arm.

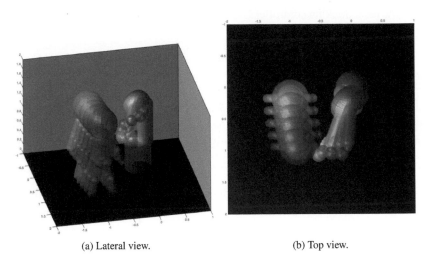

(a) Lateral view.   (b) Top view.

Figure 8.6: The simulated response of the 6DoF robot arm.

The algorithms are run in MATLAB® on a MacBook of 2.3 GHz using Intel Core i7. The running time of the safety controller is shown in Table 8.1. The average running time of the safety controller is 9.5ms, which is dominated by the time in finding the critical points, e.g., calculating the minimum distance between the robot and the human. This is because finding the critical points involves 6×10 distance calculations between capsules. If only the first three joints of the robot arm are considered, e.g., only three robot capsules are used in calculation, the running time is reduced to

Table 8.1
Running time of the safety controller.

Robot arm	Human model	Running time of the safety controller	Running time in finding critical points
6 DoF	10 capsules	9.5ms	8.8ms
3 DoF	10 capsules	5.5ms	5.0ms
6 DoF	2 capsules	2.7ms	2.0ms
3 DoF	10 spheres	0.77ms	0.40ms

5.5 ms. If the number of human capsules is reduced to two, the running time for the safety controller is reduced to 2.7 ms. Moreover, the running time of the safety controller is only 0.77 ms if the human geometry is represented using spheres. However, the spheres cannot describe the geometry as accurately as the capsules do and may be too conservative. In conclusion, current algorithms can support at least 100 Hz sampling frequency and the computation time can be further reduced if faster algorithms are developed for distance calculation.

8.2.4 DISCUSSION AND CONCLUSION

The design procedure in RSIS is summarized below.

1. Design the baseline controller that can handle the goal and the time invariant constraints.
2. Wrap every moving rigid body with a capsule to simplify the geometry.
3. Design the safe set $\mathcal{X}_S$ which specifies the required distance among capsules.
4. Design the safety index ϕ based on the safe set $\mathcal{X}_S$ and the robot dynamics.
5. Choose the control modification criteria and design the control modification metric Q.
6. Design the human motion predictor.

8.3 ROBUSTLY-SAFE AUTOMATED DRIVING SYSTEM

This section introduces the robustly-safe automated driving (ROAD) system for safe autonomous driving. The architecture of ROAD is shown in Fig. 8.7, which has three layers, corresponding to a "see-think-do" structure [124, 134] introduced earlier in Fig. 2.3. In the following discussion, section 8.3.1 introduces a multi-agent traffic model and formulates the control problem for automated driving. The details of the ROAD system will be discussed in section 8.3.2. Simulation studies will be presented in section 8.3.3. Section 8.3.4 concludes the section.

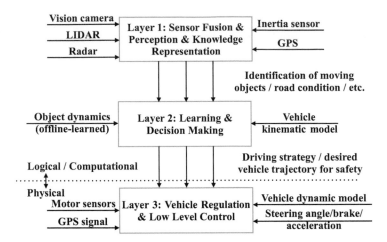

Figure 8.7: Architecture for the robustly-safe automated driving (ROAD) system.

8.3.1 MULTI-AGENT TRAFFIC MODEL

All road participants (the ego vehicle, other vehicles, and pedestrians) are viewed as agents, which have several important characteristics:

1. autonomy: the agents are self-aware and autonomous;
2. local views: no agent has a full global view of the system;
3. decentralization: there is no designated controlling agent.

From a global view, there can be thousands of agents (road participants) in the system. But only the local interactions among road participants are of interest. Hence, in the controller of the ego vehicle, only the behavior of the surrounding road participants will be analyzed. Surrounding road participants refer to the road participants that are within certain distances to the ego vehicle.

The System Model

Suppose there are N surrounding road participants locally and are indexed from 1 to N. Let $H = \{1,\dots,N\}$ be the set of indices for all surrounding road participants. The ego vehicle has index 0. Since the surrounding road participants are changing from time to time, mathematically, the topology of the multi-agent system is time varying [181].

For vehicle i, we denote its state as x_i, control input as u_i. According to (2.5), agent i's dynamic equation can be written as

$$\dot{x}_i = f_i(x_i, u_i, w_i), \tag{8.10}$$

where w_i is the disturbance introduced by the environment, e.g., wind. Let x_e be the state of the environment, e.g., speed limit v_{lim}, stationary obstacles, and so on. Then the system state x is defined as $x := [x_0^T, x_1^T, \cdots, x_N^T, x_e^T]^T$.

Agent i chooses the control u_i based on its information set π_i and its objective G_i (which can be an intended behavior or a desired tracking speed). It is assumed that there is no direct communication among road participants.[2] Agent i's information set at time T contains all the measurements up to time T, i.e., $\pi_i(T) = \{y_i(t)\}_{t \in [0,T]}$ where y_i follows from (2.3). The controller for agent i can be written as

$$u_i = g_i(\pi_i, G_i). \tag{8.11}$$

The Optimal Control Problem

In the ROAD system, the driving behavior for the ego vehicle should be designed to achieve both efficiency and safety. Equations (8.12) provide a mathematical formulation of the problem. The efficiency factor requires that the objective of the ego vehicle (such as lane following in a constant speed or going to a desired position) be achieved in an optimal manner through minimizing a cost function $J(x_0, u_0, G_0)$. As the state x_0 is not directly known, the cost should be minimized with respect to the expectation given the measurement y_0, e.g., $\mathbb{E}[J(x_0, u_0, G_0) \mid y_0]$, as shown in (8.12a). Meanwhile, the safety factor requires that the efficiency requirement be fulfilled safely. In other words, the motion of the ego vehicle should be constrained with respect to other road participants' behaviors. The constraint on the ego vehicle can be written as $x_0 \in R_S(x_1, \cdots, x_n, x_e)$ where R_S is a safe set in the state space of the ego vehicle. The set R_S depends on the states of other road participants and the state of the environment, as introduced earlier in Chapter 3. Since the states are not directly known, the constraint should be considered in the stochastic sense such that it is satisfied almost surely with probability $1 - \varepsilon$ for $\varepsilon \to 0$ given the measurement y_0, e.g., $\mathbb{P}(\{x_0 \in R_S(x_1, \cdots, x_n, x_e)\} \mid y_0) \geq 1 - \varepsilon$ as shown in (8.12e). Then the following optimal control problem is formulated

$$\min_{u_0} \mathbb{E}[J(x_0, u_0, G_0) \mid y_0], \tag{8.12a}$$

$$s.t. \ u_0 \in \Omega, \mathbb{E}(x_0 \mid y_0) \in \Gamma, \tag{8.12b}$$

$$\dot{x}_0 = f(x_0, u_0, w_0), \tag{8.12c}$$

$$y_0 = h(x_0, x_1, \cdots, x_n, x_e, v_0), \tag{8.12d}$$

$$\mathbb{P}(\{x_0 \in R_S(x_1, \cdots, x_n, x_e)\} \mid y_0) \geq 1 - \varepsilon, \tag{8.12e}$$

where Ω is the control space constraint for vehicle stability, and Γ is the constraint regarding the speed limit and other regulations. Equation (8.12c) and (8.12d) are the dynamic equation and the measurement equation respectively, where w_0 and v_0 are noise terms.

---

[2]Communication-enabled multi-vehicle navigation is discussed in section 7.4.

The ROAD system in Fig. 8.7 solves the above problem by

1. estimating the states x_i's using the measurement y_0 in Layer 1;
2. estimating the dynamics of all road participates (e.g., dynamics of x_i's) and solving for the optimal control u_0 using the estimated states in Layer 2;
3. tracking the planned trajectory in Layer 3.

8.3.2 PREDICTION AND PLANNING IN THE ROAD SYSTEM

This section focuses on Layer 2 of the ROAD system. In Layer 2, the control sequence u_0 is obtained by solving the optimization problem (8.12) given the estimated states in Layer 1. However, the optimization problem is in general hard to solve with the safety constraint (8.12e), since the dynamics of other road participants are unknown and the set R_S is non-convex. To solve the problem efficiently, the behavior design architecture discussed in section 2.2.1 will be implemented. The behavior models of other road participants will be identified and x_i's will be predicted online. A parallel planner will be used to solve the non-convex optimization efficiently.

For simplicity, we use the kinematic unicycle model for all vehicles in Layer 2. For vehicle i, let $p_i \in \mathbb{R}^2$ be the position of the center of the rear axle, $v_i \in \mathbb{R}$ the forward speed, $\theta_i \in \mathbb{R}$ the vehicle heading. Define the state $x_i = [p_i^T, v_i, \theta_i]^T$ and the control input $u_i = [\dot{v}_i, \dot{\theta}_i]^T$. Hence (8.10) can be simplified to

$$\dot{x}_i = f(x_i) + Bu_i + Bw_i, \forall i = 0, \dots, N, \tag{8.13}$$

where

$$f(x_i) = \begin{bmatrix} v_i \cos \theta_i \\ v_i \sin \theta_i \\ 0 \\ 0 \end{bmatrix}, B = \begin{bmatrix} 0 & 0 \\ 0 & 0 \\ 1 & 0 \\ 0 & 1 \end{bmatrix}.$$

In addition, define $B_1 = [0,0,1,0]^T$ and $B_2 = [0,0,0,1]^T$. Then $B = [B1, B2]$.

Cognition: Understanding Other Road Participants

Instead of predicting other vehicles' trajectories directly, human drivers may classify other drivers' intended behavior first. If the intended driving behaviors are understood, the future trajectories can be predicted using empirical models. Mimicking what humans would do, the learning structure in Fig. 8.8 is designed as an instance of the structure in Fig. 2.8 for the ego vehicle to make predictions of the surrounding vehicles, where the process is divided into two steps: (1) the behavior classification, where the observed trajectory of a vehicle goes through an offline trained classifier; and (2) the trajectory prediction, where the future trajectory is predicted based on the identified behavior, by using an empirical model which contains adjustable parameters to accommodate the driver's time-varying behavior.[3] The classification step is

---

[3] When G_i denotes vehicle i's intended behavior, the behavior classification is a backward process to identify G_i, while the trajectory prediction is a forward process to predict vehicle i's closed loop behavior $\dot{x}_i = f(x_i) + Bg_i(\pi_i, G_i)$ based on identified G_i.

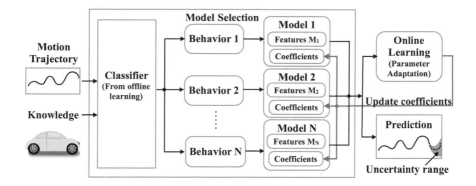

Figure 8.8: Illustration of learning and prediction in ROAD.

needed when the communications among vehicles are limited. Otherwise, vehicles can broadcast their planned behaviors as discussed in section 7.4.

The Driving Behaviors

Denote the intended behavior of vehicle i at time step k as $b_i(k)$. In this section, five behaviors are considered:

Behavior 1 ($\mathbf{B}_1$): Lane following;
Behavior 2 ($\mathbf{B}_2$): Lane changing to the left;
Behavior 3 ($\mathbf{B}_3$): Lane changing to the right;
Behavior 4 ($\mathbf{B}_4$): Lane merging;
Behavior 5 ($\mathbf{B}_5$): Lane exiting;

where $\mathbf{B}_1$ is the steady state behavior; $\mathbf{B}_2$, $\mathbf{B}_3$ and $\mathbf{B}_4$ are driving maneuvers; and $\mathbf{B}_5$ is the exiting behavior. It is assumed that there must be gaps (lane following) between two maneuvers ($\mathbf{B}_2$, $\mathbf{B}_3$ and $\mathbf{B}_4$). The transitions among behaviors follow from the model shown in Fig. 8.9a.

Let $\mathbb{P}(b_i(k) \mid \pi_0(k)) \in \mathbb{R}^5$ be the probability vector that vehicle i intends to conduct $\mathbf{B}_1, \cdots, \mathbf{B}_5$ at time step k given information up to time step k. The relationship between $\mathbb{P}(b_i(k) \mid \pi_0(k))$ and $\mathbb{P}(b_i(k+1) \mid \pi_0(k))$ can be described by a Markov matrix $A = \mathbb{P}(b_i(k+1) \mid b_i(k)) \in \mathbb{R}^{5 \times 5}$, i.e.,

$$\mathbb{P}(b_i(k+1) \mid \pi_0(k)) = A * \mathbb{P}(b_i(k) \mid \pi_0(k)), \tag{8.14}$$

where A represents the transition model[4] in Fig. 8.9a. The matrix A also corresponds to the dynamic model in (5.4).

The transition model should be invoked in calculation only when vehicle i is following the lane and is about to conduct a maneuver. When vehicle i is conducting

---

[4]A can be trained using real world labeled data $\underline{b}_i(k)$, e.g., $A_{pq} = \mathbb{P}(b_i(k+1) = \mathbf{B}_p \mid b_i(k) = \mathbf{B}_q) := \sum_{i,k} I(\underline{b}_i(k+1) = \mathbf{B}_p, \underline{b}_i(k) = \mathbf{B}_q) / \sum_{i,k} I(\underline{b}_i(k) = \mathbf{B}_q)$ where I is the indicator function.

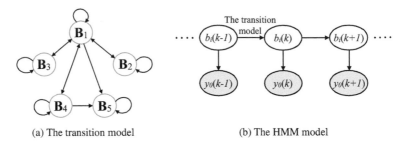

(a) The transition model   (b) The HMM model

Figure 8.9: The behavior transition model and the hidden Markov model.

a maneuver, there is no need to calculate the probability distribution over other behaviors. The transition model can be used again when the maneuver is completed
or aborted and vehicle i starts to follow the lane. The intuition is that: although the
intention of a driver is unknown (thus needs to be inferred), his action is observable
(thus when he turns his intention into action, there is no need to guess).

The Classifier and the Features

The intentions of a driver at different time steps form a Markov process as shown in
Fig. 8.9b. The Markov process is unknown to the ego vehicle. Hence the behavior
classification problem becomes an inference problem under a hidden Markov model
(HMM) [41], as discussed in section 5.2.1. According to Bayes' rule, at time step k,
for $j = 1, 2, \ldots, 5$,

$$
\begin{aligned}
\mathbb{P}(b_i(k) = \mathbf{B}_j \mid \pi_0(k)) \quad &\propto \quad \mathbb{P}(b_i(k) = \mathbf{B}_j, y_0(0), \cdots, y_0(k)), \\
&\propto \quad \mathbb{P}(y_0(k) \mid b_i(k) = \mathbf{B}_j) \mathbb{P}(b_i(k) = \mathbf{B}_j \mid \pi_0(k-1)),
\end{aligned}
$$

where $\mathbb{P}(b_i(k) \mid \pi_0(k-1))$ encodes the temporal transitions of the intended behaviors and can be obtained using (8.14). $\mathbb{P}(y_0(k) \mid b_i(k))$ is the measurement model,
which can be constructed from data offline.[5] To represent the high-dimension data
efficiently, the measurement y_0 is divided into several features for each surrounding
vehicle i (not limited to the following ones):

Feature 1: Longitudinal acceleration $f_i^1 = \dot{v}_i$.
Feature 2: Deceleration light $f_i^2 = (1, 0) = (\text{on}, \text{off})$.
Feature 3: Turn signal $f_i^3 = (1, 0, -1) = (\text{left}, \text{off}, \text{right})$.
Feature 4: Speed relative to the traffic flow $f_i^4 = v_i - \bar{v}$.
Feature 5: Speed relative to the front vehicle $f_i^5 = v_i - v_{front}$.[6]

---

[5] Similar to the transition model A, the measurement model can also be obtained by supervised training,
together with necessary curve fitting.

[6] Feature 4 and 5 encode the interactions among vehicles.

Algorithm 3 Behavior classification for vehicle i

initialization $\mathbb{P}(b_i(0)) = [1,0,0,0,0]^{\mathsf{T}}$, $k = 0$
while Classifier is Active **do**
    $k = k+1$
    read current $y_0(k)$, calculate $f_i^1(k), \cdots, f_i^{10}(k)$
    if $f_i^6(k) = -1$ **then**
        $\mathbf{B}^* = \arg\max_{\mathbf{B}=\mathbf{B}_2,\mathbf{B}_3,\mathbf{B}_4} \mathbb{P}(b_i(k-1) = \mathbf{B})$
        $\mathbb{P}(b_i(k) = \mathbf{B}^*) = 1, \mathbb{P}(b_i(k) \neq \mathbf{B}^*) = 0$
        while $f_i^6(k) = -1$ **do**
            $k = k+1$, read $y_0(k)$, calculate $f_i^6(k)$
            $\mathbb{P}(b_i(k)) = \mathbb{P}(b_i(k-1))$
        end while
        $\mathbb{P}(b_i(k)) = [1,0,0,0,0]^{\mathsf{T}}$
    else
        $\mathbb{P}(b_i(k) \mid \pi_0(k-1)) = A * \mathbb{P}(b_i(k-1) \mid \pi_0(k-1))$
        calculate $M = \mathbb{P}(y_0(k) \mid b_i(k))$
        $\mathbb{P}(b_i(k) \mid \pi_0(k)) = M . * \mathbb{P}(b_i(k) \mid \pi_0(k-1))$
        normalize $\mathbb{P}(b_i(k) \mid \pi_0(k))$
    end if
end while

Feature 6: Current lane Id f_i^6. $f_i^6 = -1$ if the vehicle is occupying two lanes.
Feature 7: Current lane clearance $f_i^7 = (1, 0, -1) = (\text{blocked}, \text{clear}, \text{ended})$.
Feature 8: Lateral velocity, e.g., $f_i^8 = v_i \sin \theta_i$.
Feature 9: Lateral deviation from the center of its current lane f_i^9.
Feature 10: Lateral deviation from the center of its target lane (if in the lane changing mode) f_i^{10}.

Algorithm 3 is designed based on the previous discussion. The probability distributions over all possible behaviors are calculated at each time step when the vehicle is following the lane (not occupying two lanes). The update of the distribution stops when the vehicle is conducting a maneuver. After the maneuver is completed, the probabilities will be initialized. An illustration of behavior classification is shown in Fig. 5.1.

The Empirical Models for Trajectory Prediction

The future trajectory is predicted according to the most likely predicted behavior, i.e., $\arg\max_{\mathbf{B}_j} \mathbb{P}(b_i(k) = \mathbf{B}_j)$. If vehicle i is following the lane, i.e., $b_i = \mathbf{B}_1$, the heading $\theta_i \approx 0$ and the lateral deviation of the vehicle can be ignored. The empirical prediction model for a lane-following vehicle assumes that the vehicle only regulates its longitudinal speed to match the speed of the traffic flow and the speed of its front vehicle, i.e.,

$$\dot{x}_i = f(x_i) + B_1[k_1 f_i^4 + k_2 f_i^5], \tag{8.15}$$

where $k_1, k_2 \in \mathbb{R}$ are adaptable parameters, which can be identified online using the method discussed in section 5.2.2. If vehicle i intents to change lane, i.e., $b_i = \mathbf{B}_2$ or $\mathbf{B}_3$, it not only regulates the longitudinal speed, but also regulates the lateral position, hence the turning rate. The empirical model for a lane-changing vehicle is designed to be

$$\dot{x}_i = f(x_i) + B_1[k_1 f_i^4 + k_2 f_i^5] + B_2[k_3 f_i^8 + k_4 f_i^{10}], \qquad (8.16)$$

where $k_1, k_2, k_3, k_4 \in \mathbb{R}$ are adaptable parameters.

Online Motion Planning and Control

Based on the predictions of the surrounding vehicles, the ego vehicle needs to find a safe and efficient trajectory satisfying the optimal control problem (8.12). The planning architecture in Fig. 8.10 is considered, as a variation of the parallel architecture in Fig. 2.3. The baseline planner solves the problem in a long time horizon without the safety constraint (8.12e), and the safety planner takes care of the safety constraint in real time. The planning architecture in Fig. 8.10 also resembles Fig. 8.2.

The Baseline Planner

The baseline planner solves the optimal control problem (8.12) to ensure efficiency, which is similar to the planner in use when the ego vehicle is navigating in an open environment. When the cost function and the control constraint are designed to be convex, (8.12) become a convex optimization problem.

Suppose the objective G_0 (target behavior and target speed v_r) is specified. The baseline planner tries to plan a trajectory to accomplish G_0. When the objective is to follow the lane, the cost function is designed as

$$J = \int_0^\infty \left[(v_0 - v_r)^2 + q(f_0^9)^2 + u_0^\mathsf{T} R u_0 \right] dt, \qquad (8.17)$$

where $q \in \mathbb{R}^+$ and $R \in \mathbb{R}^{2 \times 2}$ is positive definite. The objective penalizes the deviation from the reference speed, the deviation from the lane center, and the magnitude of the control input. When the objective is to change lane, the ego vehicle should change the lane smoothly within time T. The cost function is designed as

$$J = \int_0^T W(t) dt + S(f_0^{10}(T), \theta_0(T)), \qquad (8.18)$$

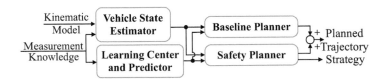

Figure 8.10: The structure of the decision making center.

where $W = (v_0 - v_r)^2 + q_1(f_0^{10})^2 + q_2(\theta_0)^2 + u_0^T R u_0$ is the run-time cost; $S = s_1 f_0^{10}(T)^2 + s_2 \theta_0(T)^2$ is the terminal cost; $q_1, q_2, s_1, s_2 \in \mathbb{R}^+$ are weights; and $R \in \mathbb{R}^{2 \times 2}$ is positive definite. The objective penalizes the deviation from the reference speed, the deviation from the center of the target lane, the heading of the vehicle, and the magnitude of the control input.

The computation in the baseline planner can be done offline. The resulting control policy will be stored for online application, to ensure real-time planning.

The Safety Planner

The safety planner modifies the trajectory planned by the baseline planner locally to ensure that it lies in the safe set $\mathcal{X}_S$ in real time as discussed in Chapter 3.

During lane following, the safety constraint requires the ego vehicle to keep a safe headway. Thus $\mathcal{X}_S(\mathbf{B}_1) = \{x : d_1(x) \geq d_{min}\}$ where $d_1(x)$ calculates the minimum distance between the ego vehicle and the vehicle or obstacles in front of it. During lane changing, the safety constraint requires the ego vehicle to keep a safe distance from vehicles on both lanes. Thus $\mathcal{X}_S(\mathbf{B}_2), \mathcal{X}_S(\mathbf{B}_3) = \{x : d_2(x) \geq d_{min}\}$ where $d_2(x)$ calculates the minimum distance between the ego vehicle and all surrounding vehicles and obstacles in the two lanes.

Mathematically, the safe set is described using a safety index ϕ, which is a real-valued continuously differentiable function on the system's state space. The state x is considered safe only if $\phi(x) \leq 0$ as shown in Fig. 3.1b. The safe region for the ego vehicle is affected by the future trajectory of the surrounding vehicles. Based on the prediction of other vehicles, if the baseline trajectory leads to $\phi \geq 0$ now or in the near future, the safety planner will generate a modification signal to decrease the safety index by making $\dot{\phi} < 0$.

The safety index is chosen as $\phi = D - d_j^2(x) - \alpha \dot{d}_j(x)$ $(j = 1, 2)$ where $D > d_{min}^2$, and $\alpha > 0$ are constants. To ensure safety, the control input of the ego vehicle must be chosen from the set of safe control $U_S(t) = \{u_0(t) : \dot{\phi} \leq -\eta_0 \text{ when } \phi \geq 0\}$ where $\eta_0 \in \mathbb{R}^+$ is a safety margin. By (8.13), the set of safe control when $\phi \geq 0$ can be written as

$$U_S(t) = \left\{ u_0(t) : \frac{\partial \phi}{\partial x_0} B u_0(t) \leq -\eta_0 - \sum_{j \in H} \frac{\partial \phi}{\partial x_j} \dot{x}_j - \frac{\partial \phi}{\partial x_0} f \right\}, \quad (8.19)$$

where $\dot{x}_j$ is the prediction made by the trajectory predictor.

If the baseline control input $u_0(t)$ is anticipated to violate the safety constraint, the safety controller will map it to the set of safe control $U_S(t)$ according to the following quadratic cost function

$$u_0^* = \min_{u \in U_S \cap \Omega} J_0(u) = \frac{1}{2}(u - u_0)^T Q(u - u_0), \quad (8.20)$$

where $Q \approx d^2 J / du_0^2$ is a positive definite matrix and defines a metric in the vehicle's control space according to the metric imposed by the cost function J in (8.17) and (8.18). In the lane following mode, if the lateral deviation f_0^9 is large due to obstacle avoidance and the safety controller continues to generate turning signal $\theta \neq 0$, the vehicle will enter the lane changing mode.

8.3.3 CASE STUDIES

Case studies are performed to illustrate the performance of the ROAD system during freeway driving. The simulations are done using parameters from a Lincoln MKZ. For freeway driving, there are typically two kinds of objectives: following a lane with a desired speed or changing lane to a target lane. The baseline planners for the two objectives are obtained offline by computing optimal control policies for (8.17) and (8.18). The safety planner checks online whether the planned trajectory is safe to execute with respect to the predicted motions of the surrounding vehicles. Three behaviors are considered for surrounding vehicles: lane following, lane change to the left, and lane change to the right. A Hidden-Markov-Model-based classifier is trained offline using labeled trajectories of human-controlled road participants in a simulator. The intended behavior of each surrounding vehicle is predicted online using the classifier. And the future motion of a surrounding vehicle is predicted using the empirical models (8.15) and (8.16) associated with the classified behaviors. In some of the simulations, the trajectory of the surrounding vehicle is manually controlled by a human subject in order to test the real time interactions.

Lane Following

Case 1—Stationary Obstacle. Figure 8.11 shows the case when the ego vehicle suddenly noticed a stationary obstacle 40 m ahead. The safety controller went active. By mapping the baseline input u_0 to U_S in (8.20), the command for deceleration and turn was generated. Then the ego vehicle slowed down and changed lane to the left to avoid the obstacle. After the lane change, the vehicle accelerated to the desired speed again.

Case 2—Slow Front Vehicle. Figure 8.12 shows the case when the front vehicle was too slow. To illustrate the interaction, the trajectories of both vehicles are down sampled and shown in the last plot in Fig. 8.12, where circles represent the ego vehicle and squares represent the slow vehicle. Different colors correspond to different time steps, the lighter the earlier. At the beginning, since it was not possible for the ego vehicle to keep the desired speed behind the slow car, it started to change lane to the left. After changing lane, it overtook the slow vehicle.

Case 3—Fast Cut-in Vehicle. Figure 8.13 shows the scenario when the ego vehicle was overtook by a fast vehicle. When the ego vehicle observed large lateral velocity from the fast vehicle, it predicted that the fast vehicle would change lane. Under the command from the safety controller, the ego vehicle slowed down. After the fast vehicle changed lane, the ego vehicle accelerated again to meet the desired speed and keep a safe headway to the fast vehicle.

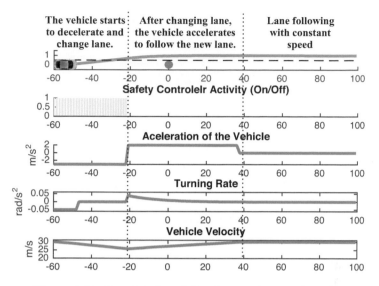

Figure 8.11: Case 1 in lane following: stationary obstacle.

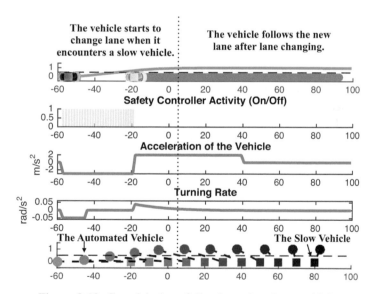

Figure 8.12: Case 2 in lane following: slow front vehicle.

Lane Change

Case 1—A Vehicle Moving Side by Side in the Target Lane. Figure 8.14 shows the case when the vehicle in the target lane was traveling next to the ego vehicle with

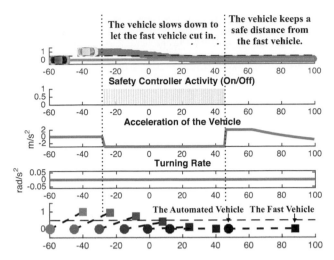

Figure 8.13: Case 3 in lane following: fast cut-in vehicle.

approximately the same speed. It was not safe to change lane in this case. Then the ego vehicle slowed down to create a gap between the two vehicles. When the distance between the two vehicles was big enough, the ego vehicle then changed to the target lane. After adjusting the relative distance to the front vehicle, the ego vehicle then followed the new lane at constant speed.

Case 2—A Slowing Down Vehicle in the Target Lane. Figure 8.15 shows the case when the vehicle in the target lane was slowing down. At first, the ego vehicle tried to use the strategy in case 1. However, when it noticed that the other vehicle also slowed down, it then sped up to overtake the other vehicle.

Case 3—Simultaneous Lane Change from Opposite Directions. Figure 8.16 shows the case when another vehicle changed to the target lane simultaneously with the ego vehicle, but from the opposite direction. At the beginning, the other vehicle was anticipated to follow its lane. Hence it was safe for the ego vehicle to change lane. When the lateral velocity of the other vehicle increased, the probability of B_3 went up and a possible future collision was anticipated. Then the safety planner went active. When the other vehicle was about to cross the lane boundary, the ego vehicle turned back to its previous lane and slowed down. The ego vehicle finally changed lane using the strategy in case 1: slowing down first and changing lane when the distance between the two vehicles was big enough.

Mixed Traffic

Figure 8.17 illustrates the active safety measures when traffic is heavy on a curved freeway. The objective G_0 is lane following with a desired speed $37 \, \mathrm{m \, s^{-1}}$ which is higher than the current traffic speed. To fulfill G_0, the vehicle performed several

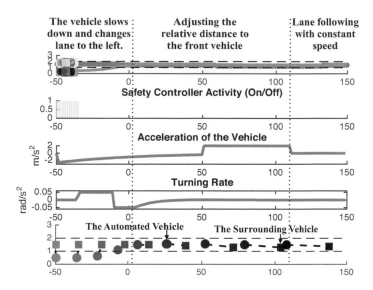

Figure 8.14: Case 1 in lane change: a parallel vehicle in the target lane.

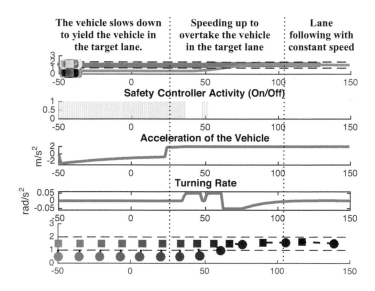

Figure 8.15: Case 2 in lane change: a slowing down vehicle in the target lane.

lane change maneuvers safely with the help of the safety planner. The distance and velocity profile during the simulation is shown in Fig. 8.17b. The dark bar indicates the moment when the safety planner was active, which matches with the moment

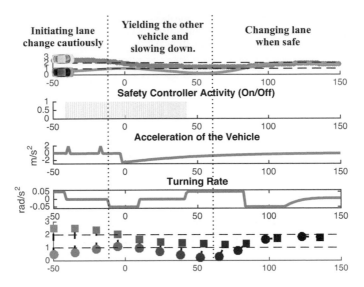

Figure 8.16: Case 3 in lane change: simultaneous lane change from opposite directions.

when the smallest distance to the surrounding vehicles reached a threshold (shown by the dotted line). The smallest distance is only computed for the surrounding vehicles that have the possibility to collide with the ego vehicle, e.g., in the moving direction of the ego vehicle. For example, if there is only one surrounding vehicle which is in the adjacent lane to the ego vehicle and both vehicles are going to follow the lanes, then the smallest distance is infinity as their moving directions are parallel to each other. When the safety planner was on, changes on the vehicle velocity and direction were generated by the safety planner as discussed earlier. The smallest distance was always kept over 4 m.

8.3.4   DISCUSSION AND CONCLUSION

The function of the ROAD system can be divided into two parts: reasoning of other road participants' behaviors and planning the trajectory for the ego vehicle. The first part relies on offline data collection and learning. The purpose of offline learning is to let the ego vehicle make reasonable predictions of the environment so that it can behave less conservatively. Even if the environment is new and the behaviors of the road participants are never encountered before, the system can still generate safe trajectories using only online learning and the parallel planners. In the beginning of the interaction, the safety planner can behave defensively by assuming the worst case scenario. During the interactions, the vehicle can fit a reactive behavior model for each road participant using the observed trajectory of that road participant, and refine the model by online adaptation, similar to the method discussed in section 5.2. The safety planner then monitors the trajectory generated by the baseline planner given the predictions made by the reactive behavior models, while a larger minimum distance requirement will be chosen if the confidence level of the model is lower. The confidence level of the model can be tracked by comparing the predicted and the observed behaviors of the corresponding road participant. Although the proposed method is mainly to address active safety for automated vehicles, it can also be applied to driving assistive systems for manually driven vehicles, if we replace the baseline planner by a function that gets human's driving command directly as shown in Fig. 8.18. Then the safety planner can monitor human's driving commands based on the predicted motions of other road participants.

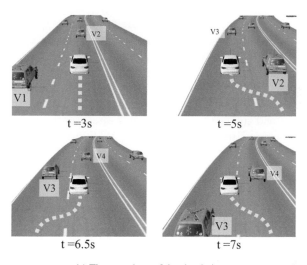

(a) The snapshots of the simulation.

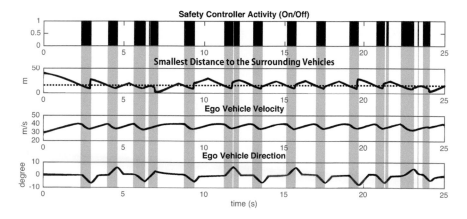

(b) The distance and velocity profile for mixed traffic simulation.

Figure 8.17: Performance of the ROAD system in heavy traffic.

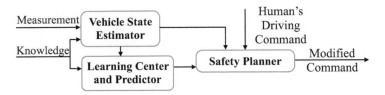

Figure 8.18: A driver assistive system using the ROAD system.

8.4 CONCLUSION

This chapter discussed techniques to enable safe and efficient space-sharing interactions among human workers and robot workers in factory floors, and among human-driven vehicles and autonomous vehicles on public roads. The two scenarios are fundamentally similar to each other. The methodologies used in the two applications are both inherited from safe control discussed in Chapter 3. Interactions among human workers and robot workers in factory floors can go beyond co-existence to hierarchical interactions and collaborations. Chapter 9 discusses hierarchical interactions in terms of robot learning from human. Chapter 10 discusses collaborations among human workers and robot workers.

9 Robot Learning from Human: Hierarchical Interactions

9.1 OVERVIEW

As discussed in Chapter 1, 'human teaching robots' is one of the most common hierarchical interactions. In this book, we have discussed imitation learning in Chapter 5 and analogy learning in Chapter 6. Both theories can serve as powerful tools to transfer the implicit skill model from a human operator (teacher) to a robot (student).

In the remainder of this chapter, the framework of remote lead through teaching will be introduced first in section 9.2 for implementing imitation learning. The applications of analogy learning, including robotic grasping and motion re-planning, will be introduced in section 9.3 and section 9.4, respectively.

9.2 REMOTE LEAD THROUGH TEACHING FOR IMPLEMENTING IMITATION LEARNING

The basic idea of *remote lead through teaching* (RLTT) is illustrated in Fig. 9.1a. A *human demonstration device* (HDD) is designed as a common tool for both human and robot. The tool frame in human demonstration phase is aligned to that in robot reproduction phase. HDD is a sensor fusion system to record all the task required information during human demonstration. While the human is performing the task, the HDD records the demonstrator's motion and force. When the robot is assigned to reproduce the task, the HDD is mounted on the robot end-effector, and sends the measurement as feedback signal to the robot controller.

The framework of RLTT is shown in Fig. 9.1b. As previously discussed, RLTT is decomposed into two phases, human demonstration and robot reproduction. In order to transfer human knowledge/skill to the robot, a skill learning process is established between the two phases.

9.2.1 HUMAN DEMONSTRATION PHASE

In the human demonstration phase, the demonstrator uses HDD to naturally perform the task. At the same time, HDD records the information in the tool frame. As mentioned previously, HDD is a combination of different sensor devices, which may

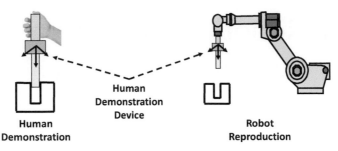

(a) Illustration of the idea.

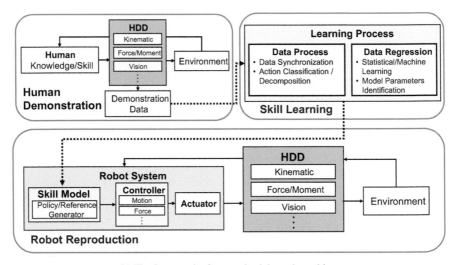

(b) The framework of remote lead through teaching.

Figure 9.1: Illustration and framework of remote lead through teaching.

include the measurement of position, velocity, and force, etc. The design of HDD can be modified according to applications.

There are several benefits of HDD in the human demonstration phase. Firstly, the human and robot workspaces are separated. Hence, the user's safety is guaranteed. Secondly, the HDD can be regarded as an add-on of the tool. It does not make significant changes of the user operating the task. Thus, the natural demonstration behavior can be preserved by RLTT. The details of the design of HDD and its data acquisition are discussed in [120].

9.2.2 SKILL LEARNING PROCESS

The purpose of skill learning process is to build the skill model from the demonstration data. The skill model is a policy or a reference generator for the robot system.

When the robot is given a feedback signal from the HDD, the skill model generates a corresponding command to the robot control loops, where the policy is learned from the human demonstration data.

Before training the model from the demonstration data, some processes are required for improving the learning quality. For instance, the demonstrator has different motion speed in each demonstration. Although the demonstration behaviors are similar, the trajectories might look very different due to the mismatched timing. Besides, robot and human have their own expertise. It is not necessary to learn every single action from human demonstration. For example, the robot moves more precisely and faster than human, while human is more intelligent in assembly. Then, the robot does not need to imitate how human approaches the workpiece, but to learn how human assembles the parts together. Hence, the data processing involves two steps. The demonstration data is firstly synchronized, then decomposed into several action segments. The data in the target segments are further utilized for skill model learning.

Since it is difficult to directly derive the human skill model, the data must be first analyzed to describe the human behavior. Statistical learning and machine learning are the methods to identify the model by the training data. If the structure of model is known, the model parameter identification technique is applicable to estimate the model parameters. We adopt the Gaussian mixture regression (GMR) technique introduced in Chapter 5 to perform skill learning, and the detail of the data process in is discussed in [119].

9.2.3 ROBOT REPRODUCTION PHASE

In the robot reproduction phase, the HDD is mounted on the robot end-effector. In addition, the skill model established from the previous process is embedded into the robot control system. Because the HDD frame in the robot reproduction phase is aligned to that in the human demonstration phase, the sensory information in these two phases are shared in the same reference. Hence, the mechanism of skill model finds the closest human demonstrated policy to the current HDD measurements.

The skill model trained by GMR is embedded into the position-force hybrid control scheme, which is shown as a shaded block in Fig. 9.2. The skill model uses the current measurements to represent Ψ, and then generates a policy as a robot reference. In this way, the human skill model could be transferred for robot applications. Note that the choice of perception will vary by tasks. For example, in the peg-hole insertion application, the force measurement is used as a feedback to generate the corrective velocity for insertion. In the surface grinding application, the current pose is used to generate the desired force.

9.2.4 EXPERIMENTAL STUDY

In order to validate the proposed RLTT framework, it is applied on two classical position-force industrial tasks. The first application scenario is the assembly task,

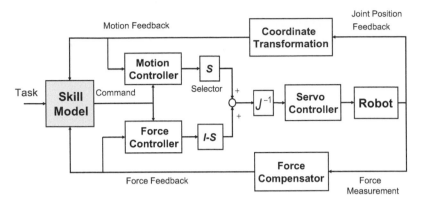

Figure 9.2: Skill model in the position-motion hybrid control scheme.

which is represented by peg-hole insertion. The second one is the grinding scenario, which is represented by a simplified testbed. Figure 9.3 shows the experimental setup for the remote lead through teaching. Figures 9.3a and b are the photos of the peg-hole insertion in human demonstration and robot reproduction phases, respectively. Figures 9.3c and d show both phases in the surface grinding. In the human demonstration phase, the demonstrator's motion is captured by tracker cameras. The robot in Figs. 9.3b and 9.3d is FANUC LR Mate 200iD/7L, where the force sensor is installed on the end-effector. However, markers are not necessary to place on the robot body because the motion of the end-effector can be obtained by calculating the forward kinematics of the current robot joint position.

Assembly Scenario

The assembly scenario is designed as an industrial standard H7/h7 peg-hole insertion task. The material and the dimension of the workpieces are listed in Table 9.1, where the tolerance is below 0.030 mm. The action segments of the insertion from the data processing are used to train the skill model. The sensed wrench is regarded as the input of the model, and the corrective velocity is regarded as the output of the model. Since the dimensions of the wrench are six, it is difficult to visualize the model. An insertion action segment is used as a query data, and the comparison of the skill model output and the original demonstration is shown in Fig. 9.4. The red-dash line is the corrective velocity from human demonstration, while the blue solid line is the velocity reference generated by the skill model. The robot has completed 50 trials in the experiment. The statistical result for the robot reproduction is shown in Table 9.2. The robot achieves 96% success rate in the H7/h7 peg-hole testbed. The two failure cases were caused by the fact that the peg was tilted before insertion.

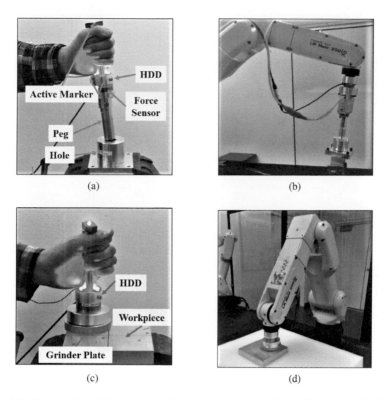

Figure 9.3: Remote lead through teaching in two industrial applications. (a) The human demonstrator performs the peg-hole insertion task. (b) The robot reproduces the peg-hole insertion experiment. (c) The human demonstrator performs the grinding task. (d) The robot reproduces the grinding experiment.

Grinding Scenario

In the grinding scenario, the demonstrator performs the grinding task as shown in Fig. 9.3c. For the safety consideration and the process simplification, an aluminum plate represents a grinder in this experiment. Also, the demonstrator only presses large forces at two specific regions of the workpiece, and moves on the surface with small force. The skill model is trained by the demonstration, where the demonstrator's pose is the input and the pressing force is the output of the model. The skill model of the demonstrator's force distribution on the workpiece is shown in Fig. 9.5a. The orange dash line in Fig. 9.5b is the force reference that sliced in the diagonal direction from Fig. 9.5a, and the blue solid line is the force that the robot applies on the workpiece along the path. The result indicates that the robot learns the human's intention in applying different forces along the trajectory.

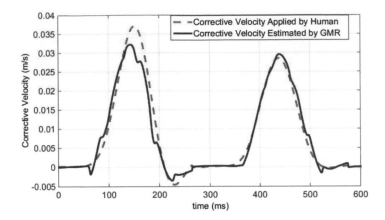

Figure 9.4: The GMR model for the assembly scenario.

Table 9.1

The specification of peg-hole insertion testbed.

	Material	Designation	Nominal Size	Upper Deviation	Lower Deviation
Hole	T6101	H7	ϕ25.400	+0.021	–0.000
Peg	T6101	h7	ϕ25.400	+0.000	–0.021

Note: Unit in millimeter (mm). Fitness type is clearance fit, and the maximum clearance is 42 μm

Table 9.2

The robot execution result in peg-hole insertion.

Total trials	Success	Fail	Success Rate(%)
50	48	2	96.00%

9.3   ROBOTIC GRASPING BY ANALOGY LEARNING

With the concept of analogy learning proposed in Chapter 6, this chapter will apply this methodology to teach industrial robots grasping skills. In general, given the point cloud of a test object, the robot will first check its similarity with all the training objects. The most similar training object will then be identified and a spatial transformation function will be constructed by structure preserved registration. With the function, the grasping pose on the training object can be transferred to achieve a new grasping pose that is suitable for the test object. Experimental results show that the proposed grasping method is efficient and robust with 94% success rate. Moreover, only a limited amount of training data is required for training the grasping skill.

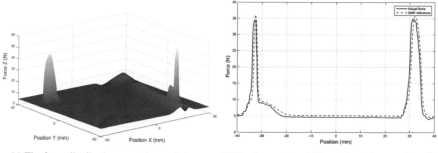

(a) The force distribution upon the workpiece.   (b) The force of the skill model and the actual applied
                                                    force.

Figure 9.5: The experiment results for the grinding scenario.

9.3.1 BACKGROUND OF ROBOTIC GRASPING

Grasping is an essential capability of robots to accomplish complex manipulation
tasks. In recent years, more and more applications require robots to grasp various objects with general purpose grippers. For example, human-robot interaction requires
collaboration and assistance between robots and humans, during which robots may
pass different tools to humans or help to hold various workpieces for assembly tasks.
The increasing demand for massive customization and warehouse automation also
promotes the development of dexterous grasping.

However, the *grasp planning* for various objects is challenging to address due
to heavy computational loads, large task variance and imperfect perceptions. Many
model-based analytic planners, such as Ferrari-Canny metric [62] and grasping
isotropy [96], require considerable time for searching and heavy computation for
evaluation. Besides, these planners generally analyze contact quality based on local
features such as contact position and contact normal, while the global task requirements such as robot reachability and collision avoidance are not under consideration. Moreover, analytic planners are usually sensitive to noise and distortions of
point clouds caused by hardware limitations and calibration errors. Therefore, the
grasping quality evaluated by analytic planners is sometimes inconsistent with the
empirical success rate and cannot resemble the reality effectively [30].

Another popular approach for grasping planning is model-free learning, i.e., learning and predicting optimal grasping from previous grasping examples. For example,
the Dex-Net [150, 151] trains a deep neural network from an enormous database
which contains 2.8 million grasping examples built by analytic planners. The network is able to estimate optimal grasping for unseen objects after training. In [223],
the grasping pose is calculated from heatmaps that were generated by a trained neural
network. These methods, however, usually require considerable data for the training
process and the optimal grasping is planned without considering the task constraints,
such as robot reachability.

Despite the variance of object shapes, we notice that objects to grasp can be classified into several categories. For example, in the tool picking scenario, objects can often be specified into categories such as wrenches, pliers, screwdrivers, etc. Objects in each category may have difference in shapes and sizes, but in general they share similar topological structures.

Some researches have been conducted based on this observation. In [36], the perceived cloud of the object is fitted to different objects, templates in the database, and the grasping is estimated by superimposing all representations considering their confidence levels. A semantic grasping is proposed in [49] to consider task requirements. The task constraints are implicitly represented by a grasping example in each object category and the desired grasping pose on the novel object is retrieved by mapping the grasping example and refined by an eigen-grasp planner. A dictionary of object parts is learned in [54] to generate grasping poses across partially similar objects. The dictionary assumes that the segments that are shared by objects are rigid and have similar sizes. However, this assumption cannot hold true in many scenarios.

In this section, we apply the analogy learning introduced in Chapter 6 to generate efficient and effective grasps from previous grasping examples. First, a series of feasible grasping poses on training objects will be demonstrated by human experts. In the test stage, the category of the test object will be classified by its similarities towards the training objects. Then a grasping pose transferring is performed between similar objects. Moreover, the transformed poses will be rated by analyzing the grasping isotropy metric [96]. An orientation search method will also be introduced to improve the robot reachability and avoid collisions.

9.3.2   GRASP PLANNING WITH HUMAN DEMONSTRATION

A general grasp planning problem with parallel grippers can be formulated as

$$\max_{\boldsymbol{c},\boldsymbol{n_c}} Q(\boldsymbol{c},\boldsymbol{n_c}) \tag{9.1a}$$

$$s.t. \quad c_i \in \partial O \quad i=1,2 \tag{9.1b}$$

$$\|c_1 - c_2\| \leq w_{\max}, \tag{9.1c}$$

where Q denotes the *grasping quality* to be maximized, $\boldsymbol{c} = \{c_1, c_2\}$ with $c_i \in \mathbb{R}^D$ denote the positions of the two contact points, and $\boldsymbol{n_c} = \{n_{c,1}, n_{c,2}\}$ denotes the normals of the contact pair with $n_{c,i} \in \mathbb{S}^{D-1}$. Constraint (9.1b) shows that the contacts should lie on the surface of object ∂O, and (9.1c) shows that the distance of the contact pair should be less than the width of the gripper $w_{\max}$.

Equation (9.1) is challenging to solve by gradient based methods because of the high complexity of surface modeling, and the discontinuity of surface presentations as well as surface normals. Compared with gradient based searching, the sampling based method is able to adapt to discrete object representation and escape from local optimum. However, it requires considerable computation for sampling and quality evaluation to find a reasonable grasp due to the complicated structure of the object and the feasibility constraints such as gripper width, task requirements and collisions, thus the direct sampling method is generally not affordable for real-time implementation.

In this chapter, we assume that the objects to grasp can be clustered into various categories. The objects in the same category share similar topological structures but can have different shapes, sizes and configurations. The objective of this work is to provide an efficient framework to grasp objects in the same category without overwhelmed training, modeling and computation. To achieve this, we introduce human demonstration to accelerate grasp searching by providing heuristics to guide sampling. Instead of directly using human demonstration as the sampling pool for the target object to grasp, we use a mapping function to transfer the example grasps based on the topological similarity between the source object and the target object. Therefore, (9.1) becomes:

$$\max_{c, n_c} Q(c, n_c) \tag{9.2a}$$

$$s.t. \quad \{c, n_c\} \in map(\mathcal{H}) \tag{9.2b}$$

$$\|c_1 - c_2\| \leq w_{\max}, \tag{9.2c}$$

where $\mathcal{H}$ denotes a human demonstration database containing example grasps on the source object, and the function $map(\cdot)$ represents a grasp transferring. Compared with (9.1), the introduction of human demonstration in (9.2) has the following advantages. First, incorporating human intelligence into the framework will improve the empirical success rate, since the human demo usually considers a variety of factors such as the local structure of the object and the global geometry for collision avoidance. Second, some tasks have special requirements. For example, some workpieces have fragile parts or polished surfaces which are not suitable for grasping. Some workpieces have some preferred grasping poses for the ease of following assembly procedures. Explicitly imposing such constraints to traditional approaches is nontrivial, while these requirements can be easily encoded by human demonstration. Moreover, by mapping the grasp examples to novel objects, the proposed method exploits much fewer but reasonable grasp samples compared to traditional exhaustive search methods. Therefore, the searching time is greatly reduced.

9.3.3   GRASPING POSE TRANSFERRING BY ANALOGY LEARNING

Assume a grasp template consists of a source object and multiple demonstrated grasping poses. A *grasping pose* is defined by the grasping position on the object and the orientation of the gripper. As Fig. 9.6 shows, the blue dots are the object point clouds and each coordinate represents a demonstrated grasping pose.

Denote the point clouds of the source object as $\mathbf{X} = (x_1, \cdots, x_N) \in \mathbb{R}^{N \times 3}$, where $x_n \in \mathbb{R}^3$ is the n-th point in the point set. The demonstrated grasping poses are denoted as $\boldsymbol{g}_i = (\boldsymbol{t}_i, \boldsymbol{R}_i) \in \mathbb{R}^3 \otimes \mathbf{SO}(3)$, $i = 1, 2, \cdots, I$, where $\boldsymbol{t}_i \in \mathbb{R}^3$ is the center of the grasping point, $\boldsymbol{R}_i \in \mathbf{SO}(3)$ represents the grasping orientation, and i is the index among the total I grasping poses. The target object is represented by another point set $\mathbf{Y} = (y_1, \cdots, y_M) \in \mathbb{R}^{M \times 3}$, where $y_m \in \mathbb{R}^3$ is the mth point in the target point set.

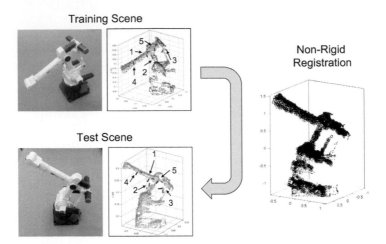

Figure 9.6: Grasping pose transferring. The point cloud of the training object is registered to the test object by non-rigid registration. A transformation function can be constructed and utilized to transfer the training grasping poses to get new ones which are feasible for the test object.

Our objective is to find a smooth transformation $\mathcal{T} : \mathbb{R}^3 \rightarrow \mathbb{R}^3$ that maps the source object to the target object as well as transferring the grasp examples to new grasping poses $g_i' = (t_i', R_i')$ on the target object (Fig. 9.6).

Grasping Pose Transferring

With the structure preserved registration (SPR) algorithm proposed in section 6.4, the desired transformation function $\mathcal{T}$ can be found by registering the two point sets. The topological structure of point sets is preserved during the alignment process so that the grasping poses can be transferred to reasonable locations.

As shown in Fig. 9.6, after finding the mapping $\mathcal{T}$, the demonstrated grasping poses on $\mathbf{X}$ will be transferred to achieve new grasping poses that are suitable for object $\mathbf{Y}$. The procedures for pose transferring can be decomposed to two parts. First, for grasping position transferring, the original grasping position can be directly mapped by function $\mathcal{T}$:

$$t_i' \leftarrow \mathcal{T}(t_i), \quad i = 1, 2, \cdots, I. \tag{9.3}$$

Second, for the grasping orientation transferring, we evaluate the Jacobian matrix of $\mathcal{T}$ (first-order partial derivatives) and utilize it to transform the original orientation to achieve a new one: $\nabla \mathcal{T}(t_i) R_i$. Singular value decomposition (SVD) is further applied to refine the orientation matrix to belong to $\mathbf{SO}(3)$ group. Specifically, the singular vectors U_i, V_i are extracted by

$$\mathbf{U}_i \Sigma \mathbf{V}_i^\mathsf{T} = svd(\nabla \mathcal{T}(\boldsymbol{t}_i)\boldsymbol{R}_i) \qquad (9.4)$$

The orientation matrix is recalculated as

$$\boldsymbol{R}_i' \leftarrow \mathbf{U}_i \mathbf{V}_i^\mathsf{T}, \quad i = 1, 2, \cdots, I. \qquad (9.5)$$

Dissimilarity Measure

During the training stage, multiple grasping poses for different categories of objects are demonstrated by human experts. Given a new object at test, it is necessary to first classify which category the object belongs to, then use structure preserved registration to transfer the corresponding grasping poses from the correct category to get a new feasible grasp. Therefore, an object classifier is essential for pose transferring.

There are some researches that apply the surface matching technique to rigidly fit the object template to the measured point clouds and calculate the *dissimilarity* [166, 97]. The source objects from different categories are exploited as the templates to match the target object. By measuring the dissimilarity between each of the source objects $\mathbf{X}$ and the target object $\mathbf{Y}$, the most similar pair will be selected to determine the category of the target object. In our work, since SPR can be applied to warp the template $\mathbf{X}$ to $\mathcal{T}(\mathbf{X})$ which is aligned with $\mathbf{Y}$, the residual dissimilarity between $\mathcal{T}(\mathbf{X})$ and $\mathbf{Y}$ instead of the dissimilarity between $\mathbf{X}$ and $\mathbf{Y}$ will be checked to provide a more robust category classification.

The average minimum distance between the two point sets can be designed as:

$$d(\mathcal{T}(\mathbf{X}), \mathbf{Y}) = \frac{1}{N} \sum_{n=1}^{N} \min_{m \in [1,M]} ||\mathcal{T}(x_n) - y_m||, \qquad (9.6)$$

where $||\mathcal{T}(x_n) - y_m||$ is the Euclidean distance between point $\mathcal{T}(x_n)$ and y_m. However, (9.6) is asymmetric. We formuate the dissimilarity between a source object and a target object to be

$$D(\mathbf{X}', \mathbf{Y}) = d(\mathbf{X}', \mathbf{Y}) + d(\mathbf{Y}, \mathbf{X}'), \qquad (9.7)$$

where $\mathbf{X}' = \mathcal{T}(\mathbf{X})$ and $D(\cdot, \cdot)$ is symmetric to its input arguments.

Suppose there are K object categories. The most possible category that the target object belongs to can be estimated by

$$\mathbf{k}^* = \arg \min_{k \in [1,K]} D(\mathbf{X}_k', \mathbf{Y}). \qquad (9.8)$$

Grasping Pose Selection

Once the object category is determined, we can map the example grasping poses from the corresponding training object to the target object.

The transformed poses serve as good candidates for grasping the test object. We will re-evaluate the quality of each transformed pose and select the best one for the robot to grasp.

The grasp quality of the transformed poses can be evaluated by analytic methods using the grasp isotropy index [96]. The grasp isotropy index measures the uniformness of different contact forces to the total wrench. More concretely, it can be written as

$$Q_i = \frac{\sigma_{\min} \mathcal{G}(\boldsymbol{g}_i', \boldsymbol{g}_o)}{\sigma_{\max} \mathcal{G}(\boldsymbol{g}_i', \boldsymbol{g}_o)}, \tag{9.9}$$

where $\boldsymbol{g}_o$ denotes the pose of the object, $\mathcal{G}(\boldsymbol{g}_i', \boldsymbol{g}_o)$ represents the grasp map determined by the contacts and the object [158], and $\sigma_{\min}$ and $\sigma_{\max}$ respectively denote the minimum and maximum singular values of the grasp map. The contacts are inferred by the line search along the grasp axis. The line search tries to locate the nearest neighbor of the grasp center on the object's point cloud. The contacts are represented by the nearest neighbors search in the positive and negative directions of the grasp axis respectively. The transferred grasp would be treated as a bad pose if the contacts deviate from the grasp axis too much, in which case a negative quality will be allocated.

Apart from the grasp quality, we also consider the feasibility constraints such as the reachability and the gripper-object collision.

A grasping example with parallel grippers is shown in Fig. 9.7a, where the blue dot is the center of grasping and the red arrows represent the grasp axis which is parallel to the operational direction of the gripper. Rotating along the grasp axis will not change the grasping quality analyzed by (9.9). However, new grasping poses can be generated, from which we can search for a valid one which is collision-free and reachable for the robots (Fig. 9.7b).

Suppose the initial orientation is denoted as $\boldsymbol{R}_0$, the sampled orientation is denoted as $\boldsymbol{R}_i$, and $\mathcal{R}$ is the set of all the sampled orientations. The orientation search can be formulated as

$$\min_{\boldsymbol{R}_i \in \mathcal{R}} \Delta\xi(\boldsymbol{R}_0, \boldsymbol{R}_i) + C\left[f_{IK}(\boldsymbol{t}, \boldsymbol{R}_i) + f_{col}(\boldsymbol{t}, \boldsymbol{R}_i, \mathbf{Y})\right], \tag{9.10}$$

where $\Delta\xi(\boldsymbol{R}_0, \boldsymbol{R}_i) = 1 - \xi(\boldsymbol{R}_0)^\mathsf{T}\xi(\boldsymbol{R}_i) \in [0,1]$ is the rotation deviation in quaternion between $\boldsymbol{R}_0$ and $\boldsymbol{R}_i$, $\xi(\cdot)$ converts a rotation matrix to a quaternion. We use quaternions rather than Euler angles to represent rotation difference to avoid singular representation in rotations. $f_{IK}(\boldsymbol{t}, \boldsymbol{R})$ is a boolean function that returns 1 when the inverse kinematics of $(\boldsymbol{t}, \boldsymbol{R})$ is invalid and returns 0 otherwise. $f_{col}(\boldsymbol{t}, \boldsymbol{R}, \mathbf{Y})$ is another boolean function which returns 1 when the gripper with pose $(\boldsymbol{t}, \boldsymbol{R}_i)$ is collided with $\mathbf{Y}$ while returns 0 otherwise. C is a large constant weight to penalize the conditions of both infeasible inverse kinematics and gripper-object collision. If all the sampled orientations are invalid, the value of (9.10) will be greater than or equal to C. Then the orientation search is applied to exploit the other candidates until it finds a feasible grasping pose to perform the task.

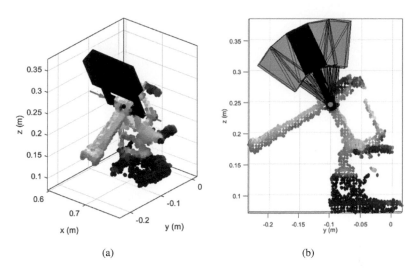

(a)   (b)

Figure 9.7: Grasping pose modification. (a) a grasp example, where the red arrows indicate the direction of gripper closing (which is also the grasp axis). (b) The side view of the grasp example. The transparent grippers share the same grasp center and grasp axis but different orientations.

9.3.4 EXPERIMENTAL STUDY

In order to verify the proposed grasping approach, a series of experiments were conducted to grasp various objects by an industrial robot manipulator. The objects were captured as point clouds from the dual 3D cameras as shown in Fig. 9.8a. By applying the snapshot of the empty workspace as a filter mask, the point clouds of objects were extracted from the background as shown in Fig. 9.8b. By running the algorithm of 'density-based spatial clustering application with noise' (DBSCAN, [60]), the point clouds can be separated to several clusters to represent different objects (Fig. 9.8c). A voxel grid filter with step size 5mm was further implemented to uniformly downsample the point clouds.

Six categories of objects, including cups, pliers, wrenches, cable adapters, toy manipulator models and toy humanoid models, were tested in the experiment (Fig. 9.9). Note that neither CAD models nor mesh files were required in this work. For each category, a specific source object was selected, and the human operator taught multiple preferred grasping poses on it by lead through teaching. The point cloud of the object and the demonstrated grasping poses were recorded as training database.

At the test stage, objects with different sizes and configurations across all the categories were randomly placed in the workspace. For example, multiple types of cups and wrenches were tested for grasping; the pliers were either open or closed; the cable adapter was twisted to various shapes; the configurations of the two different toy manipulator models and one toy humanoid model were rotated to random angles.

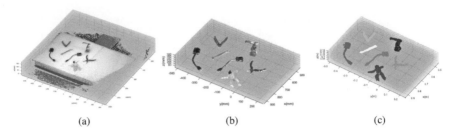

(a)   (b)   (c)

Figure 9.8: Process of point cloud. (a) The raw data was captured by the dual Ensenso stereo camera. (b) The objects were extracted from the background by predefined region. (c) The point cloud was clustered by DBSCAN.

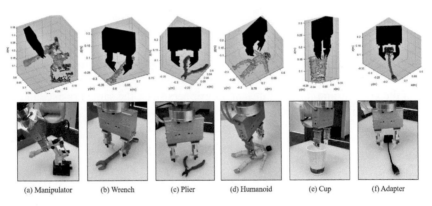

(a) Manipulator   (b) Wrench   (c) Plier   (d) Humanoid   (e) Cup   (f) Adapter

Figure 9.9: Grasp examples: the first row shows one of the grasping poses on each source object, and the second row provides the snapshots of the actual demonstrated grasping poses.

Before running the grasping experiment, an object classification test was performed by measuring the dissimilarity between the target object and all the source objects. The target objects were randomly placed, with each category of objects collecting 20 different configurations. The parameters of SPR registration were set as $\beta = 2$, $\lambda = 50$, $\mu = 0.1$, $K = 5$ and $\tau = 1e - 4$.

As shown in Fig. 9.10, the performance of object classification was presented by a confusion matrix, where each column represented the predicted class and each row represented the actual class. The diagonal entries of the confusion matrix indicated the correct classification, whereas the off-diagonal entries were misclassification. The overall classification accuracy was 94.17% (113/120). Each category of objects was tested 20 times for grasping with different orientations, shapes, sizes, and configurations. The parameters of SPR were the same as the ones in object classification.

Manipulator – 1	20	0	0	0	0	0
Wrench – 2	0	19	0	0	0	1
Plier – 3	0	2	18	0	0	0
Humanoid – 4	2	0	0	18	0	0
Cup – 5	0	0	0	0	20	0
Adapter – 6	0	2	0	0	0	18
	1	2	3	4	5	6

Figure 9.10: The confusion matrix of object classification. Each column represents a predicted class, and each row represents an actual class.

Table 9.3

Grasping quality evaluation.

Grasping Pose No.	1	2	3	4	5
Isotropy Index	0.0098	−1.000	0.0001	0.0089	−1.000

Take one target object (Fig. 9.6) as an example. The grasping qualities of the transferred grasps are provided in Table 9.3. Note that the qualities of the second and fifth transferred poses are marked as negative based on the isotropy index analysis, since the second pose was mapped to a region with sparse points, and the contacts for the fifth grasp were wider than the width of the gripper. The remaining pose with the highest grasping quality, i.e., the first pose, was selected. The selected pose was then refined by the orientation search to improve the reachability and avoid collision. The final grasp performed in the experiment is shown in Fig. 9.11b. The grasp was regarded as successful when the object could be robustly lifted up at least 10 cm without slipping. The success rate, average computation time and average point numbers for each category of objects are provided in Table 9.4. The experimental video can be found at [202]. The snapshots of grasping experiments are shown in Fig. 9.11.

Although the shapes and configurations of target objects were different from the ones of the source object, they shared similar structures. Therefore, the grasping poses on the source object could be transferred to reasonable locations on the target objects. For instance, the grasping poses on the various toy manipulator models were invariant in terms of topological structures (see the first row of Fig. 9.11). The grasping poses taught by lead through teaching had the intuition from humans such as the task specific consideration and fairly good grasping quality, and SPR transferred the insight to the target objects. Therefore, the test can be successful in most of the cases.

The failure case happened when there was a very large distortion to transform the source object to the target, which degraded the accuracy of the transformation

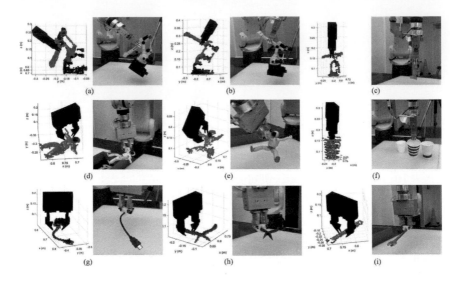

Figure 9.11: The planned grasping poses and the corresponding snapshots of the grasping results.

Table 9.4

Grasping results.

Object Class	Success / Trials	Avg. SPR Time (ms)	Avg. Numbers of Points
Manipulator	19/20	1276.4	1563.7
Wrench	20/20	111.3	316.7
Plier	18/20	706.1	1419.0
Humanoid	17/20	369.5	773.3
Cup	20/20	350.5	609.3
Adapter	19/20	480.2	917.0
Average	18.8/20	549.0	933.2

estimated by SPR. As a result, the grasping pose was not accurately transformed, which caused the grasp to fail. Although SPR did not transfer grasping poses with high accuracy in this situation, it provided a relatively close one. In the future, we may include an adaptation on the warped grasping pose to avoid this failure.

9.4   ROBOTIC MOTION RE-PLANNING BY ANALOGY LEARNING

Motion planning is one of the essential primitives for robotics. It is the foundation for manipulation, navigation and many other applications. A basic motion planning problem is to schedule a continuous motion that connects the start configuration x_S and the goal configuration x_G, while avoiding collision with obstacles, satisfying constraints (e.g., max speed, max acceleration) as well as maximizing values of rewards (e.g., smoothness, distance, control cost).

As a well-explored field in the past decades, there are already plenty of approaches developed for robotic motion planning, including grid-based [196, 84, 61, 50], sampling-based [105, 93, 92, 66] and optimization-based methods [128, 136, 177, 188].

Grid-based approaches, such as Dijkstra's algorithm [196] and A* [84], overlay a grid on the robot's configuration space, and assume each configuration is identified with a grid point. At each grid point, the robot is allowed to move to the adjacent grid points as long as it is collision-free. Heuristic functions can also be designed to accelerate the searching speed by guiding the exploration direction instead of exploring in all directions. These methods are usually complete algorithms for the path planning problem, i.e. they will find a solution eventually if it exists. However, the number of points on the grid grows exponentially in the configuration space dimension, which makes them inappropriate for high-dimensional problems, such as industrial manipulators with multiple degrees of freedom.

Instead of evaluating all possible solutions on grids, sampling-based approaches, such as probabilistic roadmap (PRM [93]) and rapidly-exploring random tree (RRT [105]), represent the configuration space with a roadmap of sampled configurations. For PRM, random samples are taken from the configuration space and connected to construct a sparse graph. The start and goal configurations are added into the graph, and a graph search algorithm is applied to determine the shortest path between the start and goal configurations. RRT creates possible paths by randomly adding points to a tree until some solution is found or the time expires. These algorithms work well for high-dimensional configuration space, because their running time is not exponentially dependent on the dimension. They are probabilistically complete, meaning that the probability that they will produce a solution approaches 1 as more time is spent.

However, the paths found by grid-based or sampling-based approaches are usually zig-zag and non-smooth. The dynamic constraints of robots are also not considered during the path search. To address these challenges, optimization-based approaches [177, 188] are proposed to formulate the motion planning task as a non-convex optimization problem, where the objective is to maximize a constructed reward function, and the constraints are initial/end conditions, system dynamics as well as configuration boundaries. The bottleneck of optimization-based planning is feasibility. Considering the non-linear robot dynamics and non-convex obstacle shapes, usually the formulated optimization for motion planning is highly non-convex. The optimization procedure might get stuck in local optimum, and sometimes even no feasible solution can be found. The selection of good initial reference becomes critical to alleviate these problems.

The aforementioned approaches achieve great success in robot motion planning. However, we observe that these approaches usually regard each motion planning

Start Point   Start Point

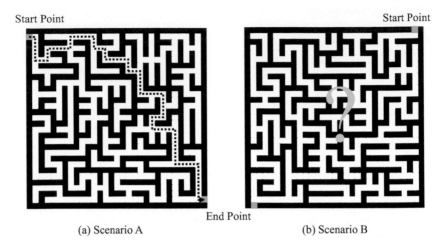

End Point

(a) Scenario A   (b) Scenario B

Figure 9.12: Find collision-free paths for robots in two mazes. Many motion planners treat the two scenarios independently and solve separately. However, scenario B is nothing but rotating scenario A with 90 degrees clockwise. Utilizing the similarity between the scenarios can achieve a feasible solution for scenario B immediately without heavy computation.

scenario as an independent problem, and seldom search for the correlation between scenarios so as to shorten the re-planning time for similar scenes.

Take the 2D motion planning problem as an example. Figure 9.12 shows two mazes with different configurations. An autonomous robot needs to plan collision-free paths under these two scenarios from the start point to the end point. Most motion planners will solve these two problems separately, i.e., planning the motion from scratch under each scenario. However, scenario B is actually a rotation of scenario A. If we simply rotate the path planned in scenario A with the same way, then we can get a feasible path immediately for B without the need of heavy computation.

With this observation, we will discuss how to apply the concept of analogy learning to accelerate robot *motion re-planning* under similar scenarios.

9.4.1 PATH RE-PLANNING BY ANALOGY LEARNING

In some robotic applications, the surrounding environment might be highly dynamic, unstructured and stochastic. It is necessary to deploy large computation power for a safe motion planning, and regard each scenario as a new one because of the low configuration repeatability. However, in many other applications, especially in industrial automation, the surrounding environment can usually be structured in advance. The configuration of the assembly line might be adjusted from time to time, so there is still need for path re-planning. However, since the configuration modification is minor, it is unnecessary to regard these configurations independently and

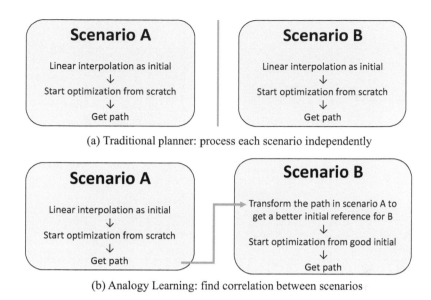

(a) Traditional planner: process each scenario independently

(b) Analogy Learning: find correlation between scenarios

Figure 9.13: Comparison between analogy learning and other approaches for path planning.

solve separately. Instead, considering the high similarities between the scenarios, we can exploit this affinity to accelerate the path planning for future similar scenarios.

Take the optimization-based path planning as an example. As shown in Fig. 9.13, for scenario A, we need to plan a 2D path for the mobile car to avoid collision with the grey obstacle. A linear interpolation between the start and end point is utilized to initialize the optimization. A collision-free path (blue line) is finally obtained by gradient descent. However, not every time a feasible solution can be achieved from the optimization solver. As discussed in section 9.4, because of the high non-convexity, the optimization process might get stuck in bad local optimum, which results in long computation time, sub-optimum result or even infeasible result. One promising method to solve this dilemma is to provide the optimization solver a 'good' *initial reference*, i.e., a reference that is as close to the global optimum as possible. Therefore, when it comes to scenario B in Fig. 9.13, a rational solver should not initialize with linear interpolation again. Instead, a spatial mapping between the two scenarios can be found, and utilized to transform the optimized path for A to get a new path for B. The transformed path then serves as the initial reference for the solver so as to boost optimization. We name the whole procedure as *path re-replanning*.

The detailed procedure of path transformation has been presented in section 6.4. Assume the environment in scenario A and B can be formulated as two point sets X and Y in the Cartesian space. The structure preserved registration (SPR, section 6.4) algorithm can be utilized to construct a non-rigid transformation $\mathcal{T} : \mathbb{R}^D \to \mathbb{R}^D$ to map X towards Y. As shown in Fig. 9.14, $\mathcal{T}$ transforms X to align to Y by warping

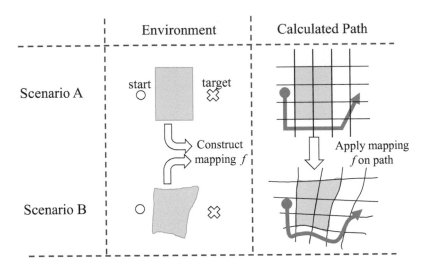

Figure 9.14: Illustration of analogy learning on path planning.

Figure 9.15: Path planning of Scenario A.

the overall configuration space. Similarly, the path P_A corresponding to scenario A can be twisted to get a new path P_B that is suitable for scenario B.

The path P of a robot can be regarded as a sequence of poses $\{p, R\}$, where p is the position, and R is the orientation. The path P_B can be achieved by applying the following transformation on P_A:

$$p_B = \mathcal{T}(p_A) \tag{9.11}$$

$$R_B = orth(J_{\mathcal{T}}(p_A) \cdot R_A) \tag{9.12}$$

$J_{\mathcal{T}}(p)$ is the Jacobian matrix of $\mathcal{T}$ evaluated at position p, and $orth(\cdot)$ is a function that orthogonalizes matrices.

A case study of path re-planning on a 6-DOF manipulator is shown in Fig. 9.15. The four purple bars are obstacles and the robot's objective is to plan a collision-free path such that its end-effector passes through the hole to reach a target point. In scenario A (Fig. 9.15), a collision-free path is planned by an optimization-based motion planner, convex feasible set (CFS, [126]). In scenario B (Fig. 9.15a), the

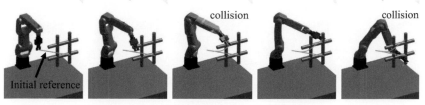

(a) Scenario B, with linear interpolation as initial reference

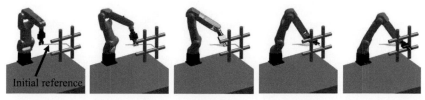

(b) Scenario B, with transformed path from Scenario A as initial reference

Figure 9.16: Path planning of Scenario B.

obstacles together with the target point shift to the left side of the robot. We still use linear interpolation between the start point and end point as the initial reference. However, CFS fails to find a collision-free path because of getting stuck into bad local optimum. In contrast, the analogy learning concept is applied to register the two obstacles and transform the path found in scenario A to get a new path for scenario B. This new path is utilized as the initial reference for CFS optimization (Fig. 9.15b). With this better initialization, an optimal and collision-free path for scenario B is found finally.

Moreover, the obstacles and target point are moved to the right side of the robot to form scenario C (Fig. 9.17a). With linear interpolation as an initial reference, CFS does succeed in finding a collision-free path for this scenario. However, the total calculation time is 3.2 sec. As a comparison, with the transformed path as initialization for optimization, CFS spends only 0.7 sec to find a collision-free solution. Note that the procedure of path transformation is also efficient, which takes 0.12 sec. The total computation speed increases 3.9 times by the proposed approach.

To conclude, path re-planning with analogy learning utilizes previous experience in past scenarios and transforms the previously solved paths to get a new one as the initial for optimization. Since this initial reference should be closer to global optimum compared to naive initialization, the optimization can benefit by avoiding bad local optimum and converging at a faster speed.

9.4.2   TRAJECTORY RE-PLANNING BY ANALOGY LEARNING

The idea of applying analogy learning for path re-planning can also be extended for *trajectory re-planning*. The only difference between the path and the trajectory is

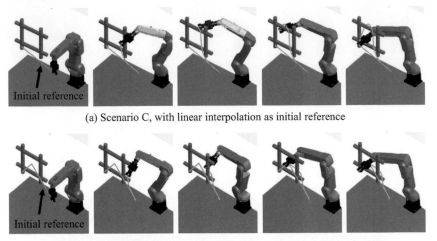

(a) Scenario C, with linear interpolation as initial reference

(b) Scenario C, with transformed path from Scenario A as initial reference

Figure 9.17: Path planning of Scenario C.

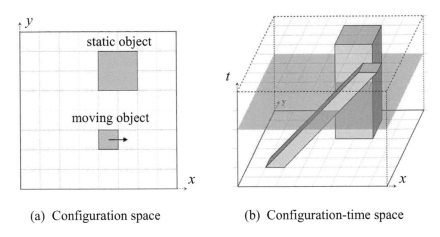

(a) Configuration space   (b) Configuration-time space

Figure 9.18: Illustration of configuration-time space.

that the trajectory has an additional time dimension, i.e., each waypoint in the path
has a corresponding requirement for pass-through time. To incorporate the dimension of time for analogy learning, the original configuration space can be augmented
and the configuration-time space is constructed as shown in Fig. 9.18. Non-rigid registration is applicable in this augmented space, and the trajectory can be transformed
as before.

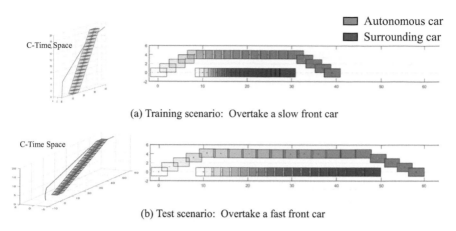

(a) Training scenario: Overtake a slow front car

(b) Test scenario: Overtake a fast front car

Figure 9.19: Trajectory replanning for test scenarios: overtaking.

Color version at the end of the book

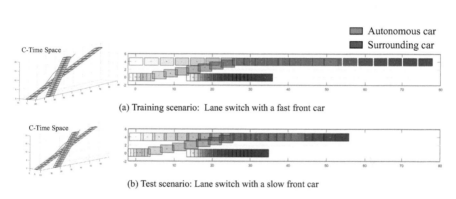

(a) Training scenario: Lane switch with a fast front car

(b) Test scenario: Lane switch with a slow front car

Figure 9.20: Trajectory replanning for test scenarios: lane switch.

Color version at the end of the book

A case study of trajectory re-planning for highway autonomous driving is shown in Fig. 9.19, Fig. 9.20 and Fig. 9.21. Three types of scenes are designed: (1) overtaking, (2) lane switching, (3) lane switch and returning back. In each case, there is a training scenario and a test scenario. The trajectory path for the training scenario is given from human demonstration. In the test scenario, since the surrounding car's speed is different, a new collision-free trajectory needs to be planned. As illustrated in the left side of Fig. 9.19, the configuration-time space is constructed for both training and test scenarios. A transformation function is found by SPR to register the two obstacles in the augmented space. The function is then utilized to transform the demonstrated trajectory (red line) to achieve a new trajectory for the test scenario. Simulation results show that the proposed trajectory replanning approach is robust to the speed change of the surrounding cars, and all the transformed trajectories all are collision-free and suitable for the test scenarios.

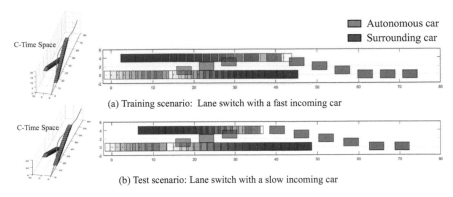

(a) Training scenario: Lane switch with a fast incoming car

(b) Test scenario: Lane switch with a slow incoming car

Figure 9.21: Trajectory replanning for test scenarios: lane switch and return back.

Color version at the end of the book

9.5 CONCLUSION

This chapter illustrated various examples of robot learning from humans.

Section 9.2 introduced a novel framework of the remote lead through teaching (RLTT) to simplify the robot programming problem. The idea of RLTT is to design a common reference between human demonstration and robot reproduction. Hence, the demonstration data from humans can be directly utilized in the robot reproduction. Two experimental verifications in the classical industrial application scenarios were given. In the assembly scenario, the robot has achieved 96% success rate in H7/h7 peg-hole insertion, where the tolerance was below 0.030 mm. In the grinding scenario, the robot successfully imitated the human behavior to press large forces in the desired spots.

Section 9.3 proposed a framework for teaching robots grasping skills based on the analogy learning approach. A database containing multiple categories of source objects together with demonstrated grasping poses were first constructed by human experts. During the test scenario, a novel object was first classified into one of the example categories by measuring its dissimilarity to each source object. Then the grasping poses on the most similar source object were transferred to the novel object by the structure preserved registration (SPR) method. The qualities of all the transferred grasping poses were evaluated and sorted by the grasp isotropy metric. The selected pose was further refined by an orientation search mechanism, which improves the robot reachability and avoids collision. A series of experiments were performed to grasp six categories of objects with various shapes, sizes and configurations. The average success rate was 18.8 out of 20 grasp trials. The experimental video is available at [202].

Section 9.4 discussed the application of analogy learning to teach robots the skill of efficient motion re-planning. Instead of regarding each scenario as an independent problem and solving them separately, we try to find the correlation between the current scenario and past solved scenarios. In this way, the past experience can be utilized and transformed to serve as a better initialization to solve the current motion planning problem. The bad local optimum for optimization can be avoided and the convergence rate can be improved. The proposed motion re-planning approach can be applied in the configuration space, configuration-time space and tangent configuration space, so that it can perform various kinds of motion planning, including path planning, trajectory planning and shape-conservative manipulation planning. A series of simulations on a 6-DOF manipulator and a 2-DOF autonomous car are implemented, which prove the effectiveness of the proposed methods.

In summary, for the force-motion hybrid skills such as assembly and grinding, the robot can imitate from human demonstration by using Gaussian mixture regression (GMR). For robotic grasping, the robot can exploit the grasp examples provided by human and efficiently generate feasible grasps by using the technique of analogy learning. Similarly, for the robot motion re-planning, the motion used in the previous scene can be transferred to the novel scene by analogy learning.

10 Human-Robot Collaboration: Time-Sharing Interactions

10.1 HUMAN-ROBOT COLLABORATION IN MANUFACTURING

This chapter focuses on human-robot collaboration in manufacturing. As discussed in Chapter 8, there are increasing interests in bringing human workers and robot workers together in production lines to leverage each others' skill and strength, e.g., humans' dexterity and flexibility and a robot's precision and productivity. For example, in BMW's factory in Spartanburg, South Carolina, robot arms cooperate with human workers to insulate and water-seal automobile doors in the final door assembly. The robot spreads out and glues down material that is held in place by the human worker's more agile hands. Before the introduction of these robots, workers had to be rotated off this uncomfortable and physically straining task after one or two hours to prevent elbow strain. In addition to cooperative robot arms, other types of cooperation are attractive [182], for example, cooperation among automated guided vehicles (AGVs) and human workers [218] in factory logistics. Such cooperation will be key to making the factories of the future flexible, productive and competitive, which will revitalize the production system and enhance the world's economy.

Introducing human-robot collaboration in factory floors poses new challenges in the design of robot behavior, especially when humans and robots are not only sharing the space, but also collaborating to finish certain tasks. A prerequisite for successful collaboration between humans and robots is to guarantee the safety of the humans. At the same time, it is important to ensure that robots collaborate with humans with the best performance possible, i.e., the robot motion should be both safe and efficient. Moreover, as production lines become more flexible, robots need to be dexterous to handle various tasks and be able to adapt to new situations. The emphasis of motion planning is shifting from rigid methods such as hard coding to flexible skill-based methods [205]. Robots should understand certain generalized skills to perform various tasks and be able to generate motion in different environments.

In well-defined and deterministic environments, state of the art safety and efficiency can be achieved. However, interactions with human workers bring a lot of uncertainties to the system. Meanwhile, the onboard computation power is limited to allow the robot to account for all possible scenarios during real time interactions. These represent major challenges faced by the co-robots. This chapter discusses methods to design the behavior of those co-robots in dynamic uncertain environments with limited computation capacity in order to maximize task efficiency while guaranteeing safety.

In the following discussion, we introduce a set of design principles of a safe and efficient robot collaboration system (SERoCS) for the next generation co-robots, which consists of robust cognition algorithms for environment monitoring, efficient task planning algorithms for safe human-robot collaboration, and safe motion planning and control algorithms for safe human-robot interactions. The proposed SE-RoCS will address the design challenges and significantly expand the skill sets of the co-robots to allow them to work safely and efficiently with their human counterparts. The development of SERoCS will create a significant advancement toward adoption of co-robots in various industries.

The remainder of this chapter is organized as follows. Section 10.2 discusses the design of SERoCS. Section 10.3 provides experimental evaluations of SERoCS. Section 10.4 concludes the chapter.

10.2 SAFE AND EFFICIENT ROBOT COLLABORATIVE SYSTEM

This section discusses the design of SERoCS, which combines various methodologies discussed in Part II. In the following discussion, the behavior design problem during human-robot collaboration is described mathematically, followed by the introduction of the SERoCS architecture that solves the problem. An example is provided to illustrate the desired performance.

10.2.1 MATHEMATICAL PROBLEM

For simplicity, this chapter focuses on the scenario with one robot and one human. The methodology extends to scenarios with multiple robots and multiple humans. Denote the robot trajectory from current time t_0 to time $t_0 + T$ as $\mathbf{x}_R = x_R(t_0 : t_0 + T)$. The planning horizon T can either be chosen as a fixed number or as a decision variable that should be optimized up to the accomplishment of the task. Similarly, the trajectories of the human and the environment from t_0 to $t_0 + T$ are $\mathbf{x}_H$ and $\mathbf{x}_e$. The sensory information π_R contains information up to current time t_0. To obtain the desired motion trajectory for the robot during human-robot collaborations, the following optimization problem is considered,

$$\min_{\mathbf{x}_R} \ \mathbb{E}\left[J(\mathbf{x}_R, \mathbf{x}_H, \mathbf{x}_e) \mid \pi_R\right], \tag{10.1a}$$

$$s.t. \ \mathbf{x}_R \in \Gamma, \tag{10.1b}$$

$$\mathbb{P}\left(\{(x_R(t), x_H(t), x_e(t)) \in \mathcal{X}_S\} \mid \pi_R\right) = 1, \forall t \in [t_0, t_0 + T], \tag{10.1c}$$

where (10.1a) is the expected cost for task performance. The cost function J evaluates the trajectories of the robot, the human, and the environment. Equation (10.1b) represents the feasibility and dynamic constraint on the robot trajectory. Equation (10.1c) is a chance constraint for safety. The safe set $\mathcal{X}_S$ is a subset of the system's state space. The system state should belong to the safe set with absolute certainty.

There are two steps in generating a desired robot motion trajectory:

1. Formulation of the problem (10.1);
2. Solving the problem (10.1) for $\mathbf{x}_R$.

During problem formulation, the trajectories $\mathbf{x}_H$ and $\mathbf{x}_e$ need to be predicted, which will be handled by an environment monitoring module. The cost function J needs to be constructed given current task progress, which will be handled by a task planning module. Finally, the optimal trajectory will be obtained by solving problem (10.1) in a motion planning module. The three modules are the main components in SERoCS, which will be discussed in detail below.

10.2.2 ARCHITECTURE OF SEROCS

The architecture of the SERoCS is shown in Fig. 10.1.

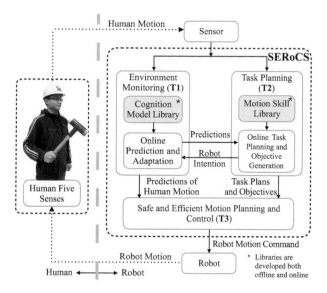

Figure 10.1: Safe and efficient robot collaboration system (SERoCS).

T1. Environment Monitoring with Human Motion Prediction

The input of this module is the sensory information π_R. The output consists of the current states $x_H(t_0)$ and $x_e(t_0)$ as well as the predicted trajectories $\hat{\mathbf{x}}_H$ and $\hat{\mathbf{x}}_e$. To generate high fidelity prediction, a learning-based method is used. Offline deep learning is employed to construct cognition models for human plan recognition and human motion prediction [45]. Online learning is designed to adapt the models to time-varying behaviors and quantify the uncertainties in the prediction as discussed in section 5.2.

T2. Task Planning with Skill Library Learned from Human Demonstration

The input of this module is the sensory information π_R and the predicted trajectories $\hat{\mathbf{x}}_H$ and $\hat{\mathbf{x}}_e$. The output is the parameters in the cost function J, especially the target pose or trajectory reference that a robot should arrive at or follow. In order to adapt to various tasks and environments, the robot learns offline to perform different tasks from human demonstration and record the knowledge in their motion skill library. During online execution, the robot adapts to different environments by generating corresponding objectives using the motion skill library as discussed in Chapter 9.

T3. Safe and Efficient Motion Planning and Control in Real Time

The input of this module is the optimization problem (10.1), where the objective function is given by task planning in T2 and the safety constraint depends on the predicted trajectories in T1. The output is the desired motion trajectory $\mathbf{x}_R$. To ensure real time computation of a feasible and safe trajectory, a parallel planning and control architecture is developed as discussed in section 2.2.1, which consists of a long-term efficiency-oriented planner and a short-term safety-oriented controller. The parallel planning and control architecture is different from the ones discussed in Chapter 8, in which the long-term planner does not consider the constraints. Real time algorithms are developed to solve the problems efficiently and make the SERoCS scalable. The algorithm for the long-term planner is discussed in Chapter 4. The algorithm for the short-term planner is discussed in Chapter 3.

10.2.3 EXAMPLE

The expected performance of SERoCS is illustrated through an example of human-robot collaborative assembly in Fig. 10.2. There are three steps in the whole process. In the first step, the robot learns the human behavior (in the example, the procedure for assembling the workpieces). In the second step, the robot learns to grasp the tool by human lead-through teaching. In the third step, the robot helps the human in finishing the assembly task which it learned in the first step by passing the tool to the human. The first two steps are called offline learning, while the last step is called online execution. The learned human behavior in the first step is recorded in the cognition model library, while the learned skill in the second step is recorded in the skill library. During online execution, the motion planning problem is formulated according to the outputs of the two libraries. The motion trajectory is then computed in real time.

10.3 EXPERIMENTAL STUDY

10.3.1 EXPERIMENT SETUP

The experiment platform is shown in Fig. 10.3. The robot manipulator is FANUC LR Mate 200iD/7L. There are one Kinect sensor to monitor the dynamic environment

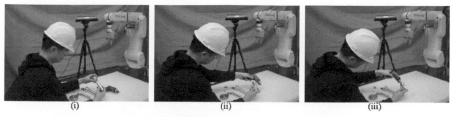

(a) Step 1: Learning the human behavior and constructing a cognition model in order to predict the human motion online (T1). (i) The human assembles the two workpieces together. (ii) The human picks up the tool. (iii) The human uses the tool to fasten the assembly.

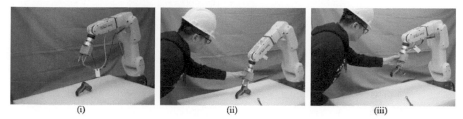

(b) Step 2: Learning to grasp the tool by lead-through-teaching (T2). (i) The configuration before teaching. (ii) The human drags the robot to the desired grasping point. (iii) The human guides the robot to grasp and lift the tool.

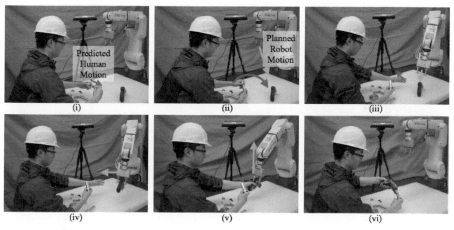

(c) Step 3: Online human-robot collaboration enabled by human motion prediction (T1) using the cognition model constructed in step 1, task planning (T2) using the learned skill in step 2, and online motion planning (T3). (i) The human assembles the two workpieces together and the robot stays away from the human. (ii) The human finishes the assembly and the robot recognizes that the human needs the tool. (iii) The robot moves toward the tool. (iv) The robot picks up the tool. (v) The robot passes the tool to the human. (vi) The human fastens the assembly using the tool and the robot steps back.

Figure 10.2: Illustration of the expected performance of the SERoCS in human-robot collaboration. The robot helps the human performing an assembly task by passing a tool. Human subjects and environment configurations vary in different steps.

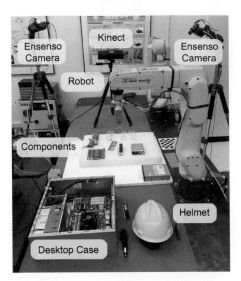

Figure 10.3: Experiment Setup.

Color version at the end of the book

and two Ensenso cameras to capture the static components placed in the workspace. For simplicity, the desktop case and the helmet are attached markers so that Kinect can directly retrieve their location in real time. All the algorithms are implemented in MATLAB® on a Windows desktop with an Intel Core i5 CPU and 16GB RAM. The robot controller is deployed on a Simulink RealTime target. Details of the experiment are elaborated in [129]. The performance of robot grasping is illustrated in section 9.3, hence omitted in this chapter.

10.3.2   VALIDATION OF THE ENVIRONMENT MONITORING

A human and a robot collaborate to assemble a desktop. The human has two plans in mind: inserting the RAMs in the motherboard first and then assembling the disk to the desktop case, or assembling the disk to the desktop case first and then inserting the RAM to the motherboard. The robot may collaborate with the human by handing the other RAM to the human if the human is assembling a RAM, or bringing the screwdriver to the human if the human is assembling the disk. To simply test the performance of T1, the robot is not plugged in in this experiment.

The trajectory of the human subject's right wrist joint is retrieved automatically from the Kinect sensor, the rate of which is about 15 frames per second. The trajectories are smoothed by a simple averaging filter. The smoothed trajectories are used to train the plan recognition classifier and the motion prediction models. 50 trajectories for each plan are collected, among which 5 are randomly chosen to be in the test set.

Under each plan, the performance of motion prediction is shown in Fig. 10.4, which aligns with the ground truth.

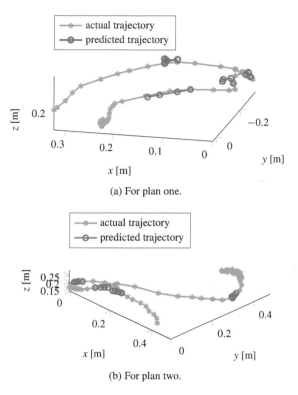

(a) For plan one.

(b) For plan two.

Figure 10.4: Human motion prediction results. The starred curves are the ground truth trajectories. Each star represents an observation at a sample time. The circled curves are the predicted trajectories at different time instances.

10.3.3   ROBOTS IN THE IDLE MODE

When the robot is in the idle mode, i.e., staying in the neutral position, it can still respond to potential dangers, as illustrated in Fig. 10.5a, Fig. 10.6a, and Fig. 10.7. In this experiment, we are using a simplified environment monitoring module. Markers are placed on the human's helmet so that the robot can track its position. In addition, a safety distance margin of 20 cm is required. Human motion is predicted using a constant speed model, which assumes that the human moves at the same speed in the near future. The Kinect runs at 10 Hz, while the safety controller runs at 1 kHz. Uncertainties are computed using a predefined maximum acceleration.

Figure 10.5a is a series of pictures taken during the experiment. The robot was in the idle mode, while the human was working on an assembly task. In the second figure, the human reached out to pick up a workpiece on the other side of the table. The human did not notice the potential collision with the end effector of the robot arm. As the robot had been actively monitoring the human movement, it moved up to yield the human. Notice that the upward movement of the robot was most efficient

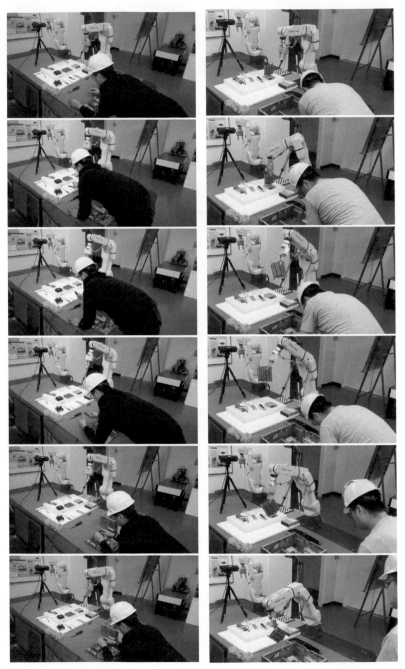

(a) Robot in the Idle Mode. (b) Human-Robot Collaboration.

Figure 10.5: Experiment of a human-robot collaboration.

Color version at the end of the book

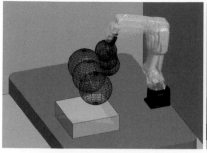 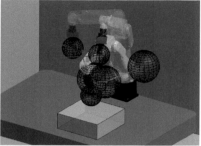

(a) Robot in the Idle Mode.   (b) Human-Robot Collaboration.

Figure 10.6: Illustration of the robot trajectory from the computation model.

Color version at the end of the book

given the prediction of the human movement. After the human got the workpiece, he went back to the sitting position. Then the robot went back to its neutral position.

In Fig. 10.6a, the red sphere represents the location of the helmet or the location of the human head. The blue sphere represents the distance margin that we enforce for the critical point. In this case, the critical point is the robot end point. The transparency of the objects (red sphere, blue sphere, and the robot arm) corresponds to different time steps, the lighter the earlier in time. The three configurations correspond to the first three figures in Fig. 10.5a. The helmet is moving towards the robot arm. To stay safe, the robot arm moves up to avoid collision.

Figure 10.7 shows the command from the safety controller, the joint velocity command sent to the robot (which includes the command from both the safety controller and the efficiency controller), the Cartesian position of the control point (in this case, the robot end point), and the minimum distance profile between the human and the robot. The shaded areas in the time axis correspond to the moment that the safety controller is in effect due to collision avoidance. The nonzero commands from the safety controller outside those shaded areas in Fig. 10.7a are due to the velocity regulation, instead of collision avoidance. There are four shaded areas. Figure 10.6a corresponds to the second shared area. The collision avoidance strategy adopted by the robot is similar in the four scenarios, that is to move the end effector up, as shown in Fig. 10.7c. The minimum distance between the human and the robot is always kept greater than 0.2 m as shown in Fig. 10.7d.

10.3.4 HUMAN-ROBOT COLLABORATIVE ASSEMBLY

The performance of SERoCS is also evaluated in a human-robot collaborative assembly task. In this task, the human was working on the cable assembly inside the desktop. Then the robot inferred that he needed to insert the motion board. As the motion board was out of reach from the human, the robot picked up the motherboard and handed it to the human. The performance of the robot is illustrated in Fig. 10.5b,

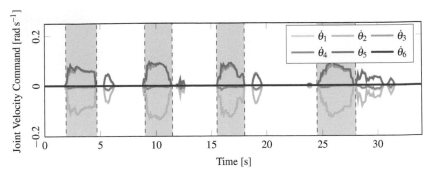

(a) The command from the safety controller.

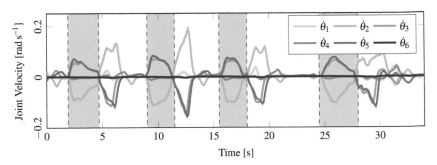

(b) The joint velocity command applied to the robot.

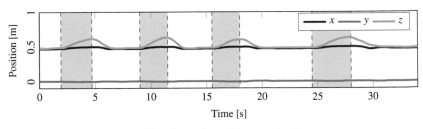

(c) The Cartesian position of the control point.

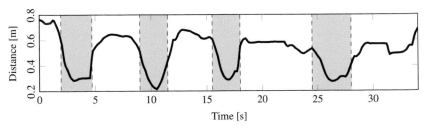

(d) The minimum distance.

Figure 10.7: Performance of the safety controller in the idle mode.

Fig. 10.6b, and Fig. 10.8. In this experiment, we are still using the simplified environment monitoring module discussed earlier. While the safety controller runs at the same sampling rate as in the previous experiment, the efficiency controller runs at 5 Hz.

Figure 10.5b is a series of pictures taken during the experiment. At the beginning, the human was doing assembly inside the desktop, while the robot decided to reach to the motherboard. The robot made the decision through human plan inference in T1. It grasped the motherboard using the skill learned in T2. While the robot was approaching the motherboard, the distance between the human and the robot was above threshold. Hence the safety controller was silent. The joint velocity command was generated by the efficiency controller, which performed online motion planning from the current position to the grasp position specified in T2. After grasping the motherboard, the robot then carried the motion board to its slot for assembly. However, as the human was too close, the robot could not directly deliver the motherboard. The efficiency controller generated a detour in order to place the workpiece from the right hand side of the human worker, which was shown in Fig. 10.6b. However, it was still not safe as the human worker was moving around. Thus, the safety controller pushed the robot arm away from the human worker. After the human finished his task inside the desktop and stayed away from the desktop, the robot inserted the motherboard using the skill learned in T2.

Figure 10.6b illustrates the configurations in the computation model. The context behind the geometric objects is the same as explained in section 10.3.3. The three configurations correspond to the third, the fourth, and the last figures in Fig. 10.5b.

Figure 10.8 shows the command from the safety controller, the joint velocity command sent to the robot (which includes the command from both the safety controller and the efficiency controller), the Cartesian position of the control point (in this case, the robot end point), and the minimum distance profile between the human and the robot. The shaded areas in the time axis correspond to the moment that the safety controller is in effect due to collision avoidance. The safety controller for collision avoidance was triggered only once. As the robot was finishing certain tasks, the joint velocity contained much richer spectrums as shown in Fig. 10.8b. The task phase of the robot can be interpreted from the location of the end effector as shown in Fig. 10.8c. The end effector both started and ended at the neutral position. The robot approached the motherboard in phase A, grasped the motherboard in phase B, moved the motherboard to the desired location (above the slot) in phase C, placed the motion board in phase D, and returned to the neutral position in phase E. Due to occlusion, there were moments that the robot lost track of the human as shown in Fig. 10.8d. In the short term (less than 1 s), the uncertainty induced by occlusion can be compensated in Task 1. However, it may put the human subject in great danger when the robot loses track of the human in a long time horizon. Avoidance of occlusion and compensation of the uncertainty induced by occlusion in the long term will be studied in the future.

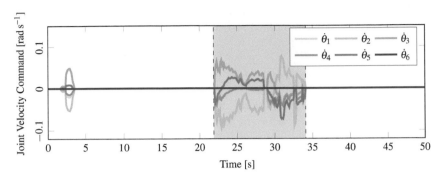

(a) The command from the safety controller.

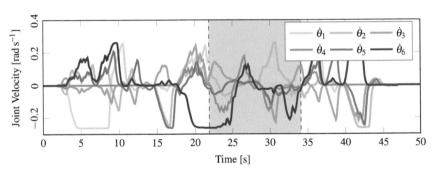

(b) The joint velocity command applied to the robot.

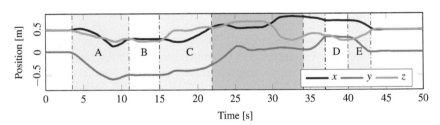

(c) The Cartesian position of the control point.

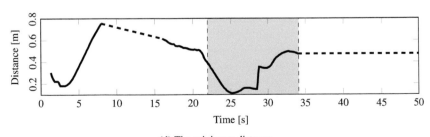

(d) The minimum distance.

Figure 10.8: Performance of SERoCS during human-robot collaboration.

10.4 CONCLUSION

This chapter discussed a set of design principles of the safe and efficient robot collaboration system (SERoCS) for the next generation co-robots, which consisted of robust cognition algorithms for environment monitoring, optimal task planning algorithms for safe human-robot collaborations, and safe motion planning and control algorithms for safe human-robot interactions. As demonstrated by the experiment, the proposed SERoCS addressed the design challenges and significantly expanded the skill sets of the co-robots to allow them to work safely and efficiently with their human counterparts. The development of SERoCS will create a significant advancement toward adoption of co-robots in various industries. SERoCS can also be applied to other time-sharing human-robot interactions.

Part IV

Conclusion

11 Vision for Future Robotics and Human-Robot Interactions

11.1 ROADMAP FOR THE FUTURE

Nowadays, robots play an increasingly important role in many application domains, from manufacturing to service, from transportation to medicare. They are expected to continuously revolutionize the industry practices and change people's life. The demands for future robots are diverse and will be addressed in future advancements.

11.1.1 THE DEMAND SIDE

The development of robotics is usually directed to address demands from various social sectors.

In manufacturing, the increasing product complexity and shorter product life-cycles introduce more difficulties to the shop floor. The current shifting from mass production to mass customization requires manufacturers to combine human dexterity and robot productivity in flexible production lines. Hence, robots should be made more intelligent to perform various complicated tasks and to safely assist human workers.

The rapid growth of the Internet of Things (IoT) and the e-commerce business revolutionizes customer behavior and expands the need of robots from factories to warehouse. For instance, online shopping replaces traditional in-store shopping and further induces the demands of automatic packaging and logistics. Unlike traditional "pick and place" tasks in conventional production lines, versatile grasping skills are required for robots to adapt to the variation among different products.

In transportation, the growing traveling demands have placed big pressure on the limited public road resources. Ride-share is a promising way to solve the conflicts. Ride-sharing with autonomous vehicles can further free human drivers and provide more space for travelers. Autonomous vehicles should be able to navigate in complicated environments and drive passengers to desired locations safely, efficiently, and comfortably.

Nevertheless, robots nowadays are not intelligent enough in both hardware and software to fulfill the aforementioned needs. Even for a simple task, it still requires tremendous hardware customization or programming efforts. Nonetheless, these demands should drive the development of future robotics.

11.1.2   FUTURE ROBOTICS

To fulfill the aforementioned demands, robots in the future should have enhanced perception, intelligence, and actuation.

Perception

A variety of sensors will be utilized to equip the robot with versatile perception skills. Robotic *vision* should be able to provide robust object recognition in real-time. The understanding of the scenes shall go beyond pixels toward meaning extraction from the scene. Moreover, the visual data shall be translated to useful information for different task requirements, e.g., location of target objects and obstacles or human's intention, etc. Robots should also be able to estimate the complicated interaction *forces* and distinguish the important force signals from external disturbances to further achieve appropriate compliance behavior. In addition to force sensing, *tactile* sensing is important for robots to distinguish texture of surface, and to exploit the force distributions to achieve more dexterous manipulation skill.

In addition to visual, force, and tactile sensing, robot communication is important for interaction with other intelligent entities. In particular, language understanding is the most natural way for smooth human-robot interactions. Future robots shall be able to understand sentences and the metaphor behind them.

To get a comprehensive understanding of the environment from all sensory information, *sensor fusion* is needed. Similar to human's five senses, robot sensor fusion comprehends multi-modality information toward a better understanding of others and the environment.

Intelligence

The key to intelligence is to be able to *optimize*, *adapt*, and *innovate*. Future robots shall always behave optimally, which is equivalent to being rational. They need to find the optimal approaches to tackle various tasks. They shall also be able to adapt to task and environment variations without human teaching or tuning. Moreover, they shall be able to generate novel behaviors from past experience instead of just mimicking or copying humans.

Actuation

On the hardware actuation side, future robots need to be *compliant* for safer interaction with humans. Compliance is also important for robots to tackle complicated tasks, such as delicate assembly. Actuation solutions should also consider mobility and sustainability. More energy-efficient actuation is needed for autonomous vehicles, mobile manipulators, and humanoids, to enable longer travel distance and operation time.

11.1.3 HUMAN-ROBOT RELATIONSHIPS

The development of robotics may fundamentally change human-robot relationships. In the past, robots only worked for humans. Today robots interact with humans as peers. The future may witness more changes. Robots may still work for humans. They can also augment humans and be like them. These aspects should be taken under consideration when we design robot behaviors.

Working for Human

Robots can work for humans in many aspects. On factory floors, robots can replace humans in dangerous and hazardous tasks, as well as in trivial, tedious, and repeated tasks. In terms of human-robot collaboration, robots can assist humans in various ways, e.g., as a helper or a process monitor.

Augmenting Human

Robots such as exoskeletons or those can be tele-operated by humans can augment human capabilities in various domains. For example, tele-immersion allows extreme environment exploration, e.g., deep sea and outer space. Medical robots augment the capability of doctors in performing delicate surgeries.

Being Human-Like

Though being human-like is not the ultimate goal of future robots, they may be like humans in order to provide emotional support to people in needs. For example, patients would prefer to have a robot nurse that they can relate to emotionally than a cold bulky machine. Moreover, as the population ages, more and more robots are needed to provide care and love to elderly people.

11.1.4 ETHICAL ISSUES

With the development of robotics, their impact on the society and economy is also worth studying and be included in design considerations. The most important ethical issue is the identification of liability and responsibility. Although researchers and practitioners work hard to ensure the safety of intelligent robots, i.e., doing no harm to the society, tragedies still happen from time to time. Meanwhile, the liability is still not clearly defined. Will the robot hold any responsibility if anything bad happens? Or will the manufacturer or the software developers hold responsibility? These are important issues to solve along the future development of robotics.

11.2 CONCLUSION OF THE BOOK

In this book, we explored the methods to design the robot behavior toward desired human-robot interactions in various scenarios. A three component behavior system was proposed, which consisted of knowledge, logic and learning. The behavior system can be designed either using model-based approaches or model-free approaches. The mathematical framework was set up in Chapter 2.

Part II of the book discussed approaches to address different design objectives: Chapter 3 for safety, Chapter 4 for efficiency, Chapter 5 for imitation, Chapter 6 for dexterity, Chapter 7 for cooperation.

Part III of the book discussed applications of the design methodologies in different scenarios: Chapter 8 for human-robot co-inhabitation, Chapter 9 for robot learning from human demonstrations, Chapter 10 for human-robot collaboration.

Design is an integrated innovative process of creating things that address functional requirements, which requires the designer to comprehensively evaluate aesthetic, psychological, functional, economic, and sociopolitical dimensions of both the design object and design process. It may involve considerable research, thought, modeling, interactive adjustment, and re-design. Unlike the discipline of mechanical design, robot behavior design is still an open area that needs coalescence. We wish this book can serve as a cornerstone for the study of robot behavior design, especially toward desirable human-robot interactions. We wish that future robots can think, behave and interact with the world in the way that human beings do, so that they can better serve, assist and collaborate with people in their daily lives across work, home and leisure.

References

1. Baxter from Rethink Robotics.
2. UR5 from Universal Robotics.
3. Kawada's NextAge robot, 2011.
4. Volvo collision avoidance features: Initial results. Technical Report 5, Highway Loss Data Institute, 2012.
5. Working with robots: Our friends electric. *The Economist*, Sep. 5, 2013. https://www.economist.com/technology-quarterly/2013/09/05/our-friends-electric, 2013.
6. Taxonomy and definitions for terms related to on-road motor vehicle automated driving systems, 2014.
7. Driverless car market watch, 2015.
8. Pieter Abbeel and Andrew Y Ng. Apprenticeship learning via inverse reinforcement learning. In *Proceedings of the International Conference on Machine learning (ICML)*, page 1, 2004.
9. Behçet Açıkmeşe, John M Carson, and Lars Blackmore. Lossless convexification of nonconvex control bound and pointing constraints of the soft landing optimal control problem. *IEEE Transactions on Control Systems Technology*, 21(6):2104–2113, 2013.
10. Nadav Aharony, Wei Pan, Cory Ip, Inas Khayal, and Alex Pentland. Social fmri: Investigating and shaping social mechanisms in the real world. *Pervasive and Mobile Computing*, 7(6):643–659, 2011.
11. Baris Akgun, Maya Cakmak, Jae Wook Yoo, and Andrea Lockerd Thomaz. Trajectories and keyframes for kinesthetic teaching: A human-robot interaction perspective. In *Proceedings of the ACM/IEEE International Conference on Human-Robot Interaction*, pages 391–398, 2012.
12. Aaron D Ames, Xiangru Xu, Jessy W Grizzle, and Paulo Tabuada. Control barrier function based quadratic programs for safety critical systems. *IEEE Transactions on Automatic Control*, 62(8):3861–3876, 2017.
13. Tanya M Anandan. Major robot OEMs fast-tracking cobots, 2014.
14. T Arai, R Kato, and M Fujita. Assessment of operator stress induced by robot collaboration in assembly. *CIRP Annals-Manufacturing Technology*, 59(1):5–8, 2010.
15. Brenna D Argall, Sonia Chernova, Manuela Veloso, and Brett Browning. A survey of robot learning from demonstration. *Robotics and Autonomous Systems*, 57(5):469–483, 2009.
16. Christopher G Atkeson, Eric W Aboaf, Joseph McIntyre, and David J Reinkensmeyer. Model-based robot learning. 1988.
17. Christopher G Atkeson and Stefan Schaal. Robot learning from demonstration. In *ICML*, volume 97, pages 12–20, 1997.
18. Seyed Azimi, Gaurav Bhatia, Ragunathan Rajkumar, and Priyantha Mudalige. Reliable intersection protocols using vehicular networks. In *2013 ACM/IEEE International Conference on Cyber-Physical Systems*, pages 1–10.
19. Joonbum Bae and Masayoshi Tomizuka. Gait phase analysis based on a hidden Markov model. *IEEE/ASME Transactions on Mechatronics*, 21(6):961–970, 2011.
20. Lisanne Bainbridge. Ironies of automation. *Automatica*, 19(6):775–779, 1983.
21. Albert Bandura. Social cognitive theory: An agentic perspective. *Annual Review of Psychology*, 52(1):1–26, 2001.

22. Tamer Basar and Geert Jan Olsder. *Dynamic Noncooperative Game Theory*, volume 200. London: Academic Press, 1995.

23. George A Bekey. *Autonomous Robots: From Biological Inspiration to Implementation and Control*. MIT Press, 2005.

24. Darrin C Bentivegna, Christopher G Atkeson, and Gordon Cheng. Learning tasks from observation and practice. *Robotics and Autonomous Systems*, 47(2-3):163–169, 2004.

25. Charles R Berger and Richard J Calabrese. Some explorations in initial interaction and beyond: Toward a developmental theory of interpersonal communication. *Human Communication Research*, 1(2):99–112, 1975.

26. Dimitris Bertsimas, Vishal Gupta, and Ioannis Ch Paschalidis. Data-driven estimation in equilibrium using inverse optimization. *Mathematical Programming*, pages 1–39, 2014.

27. Jeff A Bilmes et al. A gentle tutorial of the em algorithm and its application to parameter estimation for gaussian mixture and hidden markov models. *International Computer Science Institute*, 4(510):126, 1998.

28. Christopher M Bishop. *Pattern recognition and machine learning*, volume 1. Springer: New York, 2006.

29. Paul T Boggs and Jon W Tolle. Sequential quadratic programming. *Acta Numerica*, 4:1–51, 1995.

30. Jeannette Bohg, Antonio Morales, Tamim Asfour, and Danica Kragic. Data-driven grasp synthesis—a survey. *IEEE Transactions on Robotics*, 30(2):289–309, 2014.

31. Mariusz Bojarski, Davide Del Testa, Daniel Dworakowski, Bernhard Firner, Beat Flepp, Prasoon Goyal, Lawrence D Jackel, Mathew Monfort, Urs Muller, Jiakai Zhang, et al. End to end learning for self-driving cars. *arXiv preprint arXiv:1604.07316*, 2016.

32. Paulo Vinicius Koerich Borges, Nicola Conci, and Andrea Cavallaro. Video-based human behavior understanding: a survey. *IEEE Transactions on Circuits and Systems for Video Technology*, 23(11):1993–2008, 2013.

33. S Bouchard. With two arms and a smile, Pi4 Workerbot is one happy factory bot. *IEEE Spectrum*, 2011.

34. Cynthia Breazeal. Social interactions in hri: the robot view. *IEEE Transactions on Systems, Man, and Cybernetics-Part C: Applications and Reviews*, 34(2):181–186, 2004.

35. Douglas A Bristow, Marina Tharayil, and Andrew G Alleyne. A survey of iterative learning control. *IEEE Control Systems*, 26(3):96–114, 2006.

36. Peter Brook, Matei Ciocarlie, and Kaijen Hsiao. Collaborative grasp planning with multiple object representations. In *Robotics and Automation (ICRA), 2011 IEEE International Conference on*, pages 2851–2858. IEEE, 2011.

37. Lawrence D Burns. Sustainable mobility: a vision of our transport future. *Nature*, 497(7448):181–182, 2013.

38. Richard H Byrd, Jorge Nocedal, and Richard A. Waltz. pages 35–59. Springer US, Boston, MA.

39. S Calinon and A Billard. Learning of Gestures by Imitation in a Humanoid Robot. In *Imitation and Social Learning in Robots, Humans and Animals: Behavioural, Social and Communicative Dimensions*, pages 153–177. Cambridge University Press, 2007. K Dautenhahn and CL Nehaniv (Eds).

40. Sylvain Calinon, Florent D'halluin, Eric L Sauser, Darwin G Caldwell, and Aude G Billard. Learning and reproduction of gestures by imitation. *IEEE Robotics and Automation Magazine*, 17(2):44–54, 2010.

41. Olivier Cappé, Eric Moulines, and Tobias Rydén. *Inference in Hidden Markov Models*. Springer Science & Business Media, New York, 2006.

42. Jaime G Carbonell. Learning by analogy: Formulating and generalizing plans from past experience. In *Machine Learning, Volume I*, pages 137–161. Springer, Berlin, Heidelberg, 1983.

43. George Charalambous, Sarah Fletcher, and Philip Webb. Human-automation collaboration in manufacturing. In Contemporary Ergonomics and Human Factors 2013, volume 59, pages 59–66. ROUTLEDGE in association with GSE Research, 2013.

44. Yutuo Chen, Xuli Han, Minoru Okada, Yu Chen, and Fazel Naghdy. Intelligent robotic peg-in-hole insertion learning based on haptic virtual environment. In *Computer-Aided Design and Computer Graphics, 2007 10th IEEE International Conference on*, pages 355–360. IEEE, 2007.

45. Yujiao Cheng, Weiye Zhao, Changliu Liu, and Masayoshi Tomizuka. Human motion prediction using adaptable neural networks. *arXiv:1810.00781*, 2018.

46. Rahul Chipalkatty, Hannes Daepp, Magnus Egerstedt, and Wayne Book. Human-in-the-loop: Mpc for shared control of a quadruped rescue robot. In *Proceedings of the IEEE/RSJ International Conference on Intelligent Robots and Systems (IROS)*, pages 4556–4561, 2011.

47. Haili Chui and Anand Rangarajan. A new point matching algorithm for non-rigid registration. *Computer Vision and Image Understanding*, 89(2):114–141, 2003.

48. ChuiDataset. Data set for testing non-rigid registration. `http://legacydirs.umiacs.umd.edu/\%7Ezhengyf/PointMatching`, 2000.

49. Hao Dang and Peter K Allen. Semantic grasping: Planning robotic grasps functionally suitable for an object manipulation task. In *Intelligent Robots and Systems (IROS), 2012 IEEE/RSJ International Conference on*, pages 1311–1317. IEEE, 2012.

50. Kenny Daniel, Alex Nash, Sven Koenig, and Ariel Felner. Theta*: Any-angle path planning on grids. *Journal of Artificial Intelligence Research*, 39:533–579, 2010.

51. Kerstin Dautenhahn. Robots we like to live with?!– a developmental perspective on a personalized, life-long robot companion. In *Proceedings of the IEEE International Workshop on Robot and Human Interactive Communication*, pages 17–22, 2004.

52. Kerstin Dautenhahn. Socially intelligent robots: Dimensions of human–robot interaction. *Philosophical Transactions of the Royal Society B: Biological Sciences*, 362(1480):679–704, 2007.

53. Arthur P Dempster, Nan M Laird, and Donald B Rubin. Maximum likelihood from incomplete data via the em algorithm. *Journal of the Royal Statistical Society. Series B (Methodological)*, pages 1–38, 1977.

54. Renaud Detry, Carl Henrik Ek, Marianna Madry, Justus Piater, and Danica Kragic. Generalizing grasps across partly similar objects. In *Robotics and Automation (ICRA), 2012 IEEE International Conference on*, pages 3791–3797. IEEE, 2012.

55. M Bernardine Dias, Balajee Kannan, Brett Browning, E Gil Jones, Brenna Argall, M Freddie Dias, Marc Zinck, Manuela M Veloso, and Anthony J Stentz. Sliding autonomy for peer-to-peer human-robot teams. In *Proceedings of the Intelligent Conference on Intelligent Autonomous Systems (IAS)*, pages 332–341, 2008.

56. Noel E Du Toit and Joel W Burdick. Robot motion planning in dynamic, uncertain environments. *IEEE Transactions on Robotics*, 28(1):101–115, 2012.

57. David Eberly. *Robust Computation of Distance Between Line Segments*, 2015.

58. Gabriele Eichfelder and Janez Povh. On the set-semidefinite representation of nonconvex quadratic programs over arbitrary feasible sets. *Optimization Letters*, 7(6):1373–1386, 2013.

59. M Elhenawy, AA Elbery, AA Hassan, and HA Rakha. An intersection game-theory-based traffic control algorithm in a connected vehicle environment. In *2015 IEEE International Conference on Intelligent Transportation Systems*, pages 343–347.

60. Martin Ester, Hans-Peter Kriegel, Jörg Sander, Xiaowei Xu, et al. A density-based algorithm for discovering clusters in large spatial databases with noise. In *Kdd*, volume 96, pages 226–231, 1996.

61. Dave Ferguson and Anthony Stentz. Field d*: An interpolation-based path planner and replanner. In *Robotics Research*, pages 239–253. Springer, 2007.

62. Carlo Ferrari and John Canny. Planning optimal grasps. In *Robotics and Automation, 1992. Proceedings, 1992 IEEE International Conference on*, pages 2290–2295. IEEE, 1992.

63. Terrence Fong, Illah Nourbakhsh, and Kerstin Dautenhahn. A survey of socially interactive robots. *Robotics and Autonomous Systems*, 42(3):143–166, 2003.

64. Katerina Fragkiadaki, Sergey Levine, Panna Felsen, and Jitendra Malik. Recurrent network models for human dynamics. In *Proceedings of the IEEE International Conference on Computer Vision (ICCV)*, pages 4346–4354, 2015.

65. Antonio Franchi, Cristian Secchi, Hyoung Il Son, Heinrich H Bülthoff, and Paolo Robuffo Giordano. Bilateral teleoperation of groups of mobile robots with time-varying topology. *IEEE Transactions on Robotics*, 28(5):1019–1033, 2012.

66. Jonathan D Gammell, Siddhartha S Srinivasa, and Timothy D Barfoot. Informed rrt*: Optimal sampling-based path planning focused via direct sampling of an admissible ellipsoidal heuristic. In *Intelligent Robots and Systems (IROS 2014), 2014 IEEE/RSJ International Conference on*, pages 2997–3004. IEEE, 2014.

67. Song Ge, Guoliang Fan, and Meng Ding. Non-rigid point set registration with global-local topology preservation. In *Computer Vision and Pattern Recognition Workshops (CVPRW), 2014 IEEE Conference on*, pages 245–251. IEEE, 2014.

68. Jeremy H Gillula, Haomiao Huang, Michael P Vitus, and Claire J Tomlin. Design of guaranteed safe maneuvers using reachable sets: Autonomous quadrotor aerobatics in theory and practice. In *2010 IEEE International Conference on Robotics and Automation*, pages 1649–1654. IEEE, 2010.

69. Federico Girosi, Michael Jones, and Tomaso Poggio. Regularization theory and neural networks architectures. *Neural Computation*, 7(2):219–269, 1995.

70. Matthew Golub, Steven Chase, and M Yu Byron. Learning an internal dynamics model from control demonstration. In *Proceedings of the International Conference on Machine Learning (ICML)*, pages 606–614, 2013.

71. D González, J Pérez, V Milanés, and F Nashashibi. A review of motion planning techniques for automated vehicles. *IEEE Transactions on Intelligent Transportation Systems*, 17(4):1135–1145, 2016.

72. Michael A Goodrich and Alan C Schultz. Human-robot interaction: A survey. *Foundations and Trends in Human-Computer Interaction*, 1(3):203–275, 2007.

73. Graham C Goodwin and Kwai Sang Sin. *Adaptive Filtering Prediction and Control*. Courier Dover Publications, 2013.

74. Luis Gracia, Fabricio Garelli, and Antonio Sala. Reactive siding-mode algorithm for collision avoidance in robotic systems. *IEEE Transactions on Control Systems Technology*, 21(6):2391–2399, 2013.

75. Leslie Greengard and John Strain. The fast gauss transform. *SIAM Journal on Scientific and Statistical Computing*, 12(1):79–94, 1991.

76. Elena Gribovskaya, Abderrahmane Kheddar, and Aude Billard. Motion learning and adaptive impedance for robot control during physical interaction with humans. In *Proceedings of the IEEE International Conference on Robotics and Automation (ICRA)*, pages 4326–4332, 2011.

77. Tianyu Gu, J Atwood, Chiyu Dong, JM Dolan, and Jin-Woo Lee. Tunable and stable real-time trajectory planning for urban autonomous driving. In *Proceedings of the IEEE/RSJ International Conference on Intelligent Robots and Systems (IROS)*, pages 250–256, 2015.

78. Tianyu Gu, John M Dolan, and Jin-Woo Lee. Runtime-bounded tunable motion planning for autonomous driving. In *Proceedings of the IEEE Intelligent Vehicles Symposium (IV)*, pages 1301–1306, 2016.

79. Sami Haddadin, A Albu-Schaffer, Oliver Eiberger, and Gerd Hirzinger. New insights concerning intrinsic joint elasticity for safety. In *Proceedings of the IEEE/RSJ International Conference on Intelligent Robots and Systems (IROS)*, pages 2181–2187, 2010.

80. Sami Haddadin, Alin Albu-Schaffer, Alessandro De Luca, and Gerd Hirzinger. Collision detection and reaction: A contribution to safe physical human-robot interaction. In *Proceedings of the IEEE/RSJ International Conference on Intelligent Robots and Systems (IROS)*, pages 3356–3363, 2008.

81. Sami Haddadin, Michael Suppa, Stefan Fuchs, Tim Bodenmüller, Alin Albu-Schäffer, and Gerd Hirzinger. Towards the robotic co-worker. In *Robotics Research*, volume 70, pages 261–282. Springer Berlin Heidelberg, 2011.

82. Christopher Harper and Gurvinder Virk. Towards the development of international safety standards for human robot interaction. *International Journal of Social Robotics*, 2(3):229–234, 2010.

83. Matthew W Harris and Behçet Açıkmeşe. Lossless convexification of non-convex optimal control problems for state constrained linear systems. *Automatica*, 50(9):2304–2311, 2014.

84. Peter E Hart, Nils J Nilsson, and Bertram Raphael. A formal basis for the heuristic determination of minimum cost paths. *IEEE Transactions on Systems Science and Cybernetics*, 4(2):100–107, 1968.

85. John A Hartigan and Manchek A Wong. Algorithm as 136: A k-means clustering algorithm. *Applied Statistics*, pages 100–108, 1979.

86. Richard F Hartl, Suresh P Sethi, and Raymond G Vickson. A survey of the maximum principles for optimal control problems with state constraints. *SIAM Review*, 37(2):181–218, 1995.

87. Nicholas J Higham. *Analysis of the Cholesky decomposition of a semi-definite matrix*. Oxford University Press, 1990.

88. Gerd Hirzinger, A Albu-Schaffer, M Hahnle, Ingo Schaefer, and Norbert Sporer. On a new generation of torque controlled light-weight robots. In *Proceedings of the IEEE International Conference on Robotics and Automation (ICRA)*, volume 4, pages 3356–3363, 2001.

89. Robert Hof. Toyota: 'guardian angel' cars will beat self-driving cars. *Forbes*, 2016.

90. Thomas M Howard, Colin J Green, and Alonzo Kelly. Receding horizon model-predictive control for mobile robot navigation of intricate paths. In *Field and Service Robotics*, pages 69–78, 2010.

91. Tor Arne Johansen, Thor I Fossen, and Stig P Berge. Constrained nonlinear control allocation with singularity avoidance using sequential quadratic programming. *IEEE Transactions on Control Systems Technology*, 12(1):211–216, 2004.

92. Sertac Karaman and Emilio Frazzoli. Incremental sampling-based algorithms for optimal motion planning. *Robotics Science and Systems VI*, 104:2, 2010.

93. Lydia E Kavraki, Petr Svestka, J-C Latombe, and Mark H Overmars. Probabilistic roadmaps for path planning in high-dimensional configuration spaces. *IEEE Transactions on Robotics and Automation*, 12(4):566–580, 1996.

94. S Mohammad Khansari-Zadeh, Klas Kronander, and Aude Billard. Modeling robot discrete movements with state-varying stiffness and damping: A framework for integrated motion generation and impedance control. *Proceedings of Robotics: Science and Systems X (RSS 2014)*, 2014.

95. Oussama Khatib. Real-time obstacle avoidance for manipulators and mobile robots. *The International Journal of Robotics Research*, 5(1):90–98, 1986.

96. Byoung-Ho Kim, Sang-Rok Oh, Byung-Ju Yi, and Il Hong Suh. Optimal grasping based on non-dimensionalized performance indices. In *Intelligent Robots and Systems, 2001. Proceedings. 2001 IEEE/RSJ International Conference on*, volume 2, pages 949–956. IEEE, 2001.

97. Ulrich Klank, Dejan Pangercic, Radu Bogdan Rusu, and Michael Beetz. Real-time cad model matching for mobile manipulation and grasping. In *Humanoid Robots, 2009. Humanoids 2009. 9th IEEE-RAS International Conference on*, pages 290–296. IEEE, 2009.

98. Jens Kober and Jan Peters. Reinforcement learning in robotics: A survey. In *Reinforcement Learning*, pages 579–610. Springer, 2012.

99. Thomas Kollar and Nicholas Roy. Trajectory optimization using reinforcement learning for map exploration. *The International Journal of Robotics Research*, 27(2):175–196, 2008.

100. Kyoungchul Kong, Joonbum Bae, and Masayoshi Tomizuka. Control of rotary series elastic actuator for ideal force-mode actuation in human–robot interaction applications. *IEEE/ASME Transactions on Mechatronics*, 14(1):105–118, 2009.

101. Klas Kronander, Etienne Burdet, and Aude Billard. Task transfer via collaborative manipulation for insertion assembly. In *Workshop on Human-Robot Interaction for Industrial Manufacturing, Robotics, Science and Systems*. Citeseer, 2014.

102. Jörg Krüger, Terje K Lien, and Alexander Verl. Cooperation of human and machines in assembly lines. *CIRP Annals-Manufacturing Technology*, 58(2):628–646, 2009.

103. Marek Kuczma. *An introduction to the theory of functional equations and inequalities: Cauchy's equation and Jensen's inequality*. Springer Science & Business Media, 2009.

104. Alex Kuefler, Jeremy Morton, Tim A Wheeler, and Mykel J Kochenderfer. Imitating driver behavior with generative adversarial networks. In *Proceedings of the IEEE Intelligent Vehicles Symposium (IV)*, 2017.

105. James J Kuffner and Steven M LaValle. RRT-connect: An efficient approach to single-query path planning. In *Proceedings of the IEEE International Conference on Robotics and Automation (ICRA)*, volume 2, pages 995–1001, 2000.

106. Joseph P La Salle. An invariance principle in the theory of stability. 1966.

107. Ioan Doré Landau, Rogelio Lozano, and Mohammed M'Saad. *Adaptive control*, volume 51. Springer, 1998.

108. Steven M LaValle and James J Kuffner Jr. Rapidly-exploring random trees: Progress and prospects. In *Algorithmic and Computational Robotics: New Directions*, pages 293–308, 2000.

109. Jessica Leber. At volkswagen, robots are coming out of their cages, 2013.

110. Yann LeCun, Yoshua Bengio, and Geoffrey Hinton. Deep learning. *Nature*, 521(7553):436, 2015.

111. Insuk Lee, Eiru Kim, and Edward M Marcotte. Modes of interaction between individuals dominate the topologies of real world networks. *PLOS ONE*, 10(3):1–12, 2015.

112. Ian Lenz, Honglak Lee, and Ashutosh Saxena. Deep learning for detecting robotic grasps. *The International Journal of Robotics Research*, 34(4-5):705–724, 2015.

113. Sergey Levine, Chelsea Finn, Trevor Darrell, and Pieter Abbeel. End-to-end training of deep visuomotor policies. *Journal of Machine Learning Research*, 17(39):1–40, 2016.

114. Sergey Levine, Nolan Wagener, and Pieter Abbeel. Learning contact-rich manipulation skills with guided policy search. In *Robotics and Automation (ICRA), 2015 IEEE International Conference on*, pages 156–163. IEEE, 2015.

115. X Li, Z Sun, Z He, Q Zhu, and D Liu. A practical trajectory planning framework for autonomous ground vehicles driving in urban environments. In *Proceedings of the IEEE Intelligent Vehicles Symposium (IV)*, pages 1160–1166, 2015.

116. Chu-Min Liao and Richard SW Masters. Analogy learning: A means to implicit motor learning. *Journal of Sports Sciences*, 19(5):307–319, 2001.

117. Hsien-Chung Lin, Changliu Liu, and Masayoshi Tomizuka. Fast robot motion planning with collision avoidance and temporal optimization. In *2018 International Conference on Control, Automation, Robotics and Vision (ICARCV)*, pages 29–35. IEEE, 2018.

118. Hsien-Chung Lin, Te Tang, Yongxiang Fan, and Masayoshi Tomizuka. A framework for robot grasp transferring with non-rigid transformation. In *2018 IEEE/RSJ International Conference on Intelligent Robots and Systems (IROS)*, pages 2941–2948, Oct 2018.

119. Hsien-Chung Lin, Te Tang, Yongxiang Fan, Yu Zhao, Masayoshi Tomizuka, and Wenjie Chen. Robot learning from human demonstration with remote lead through teaching. In *Proceedings of the European Control Conference (ECC)*, 2016.

120. Hsien-Chung Lin, Te Tang, Masayoshi Tomizuka, and Wenjie Chen. Remote lead through teaching by human demonstration device. In *Proceedings of the ASME Dynamic Systems and Control Conference (DSCC)*, pages V002T30A003–V002T30A003, 2015.

121. Michael L Littman. Markov games as a framework for multi-agent reinforcement learning. In *Proceedings of the International Conference on Machine Learning (ICML)*, volume 157, pages 157–163, 1994.

122. Chang Liu, S Lee, S Varnhagen, and HE Tseng. Path planning for autonomous vehicles using model predictive control. In *Proceedings of the IEEE Intelligent Vehicles Symposium (IV)*, pages 174–179, 2017.

123. Changliu Liu. Real time robot motion planning in dynamic uncertain environment.

124. Changliu Liu, Jianyu Chen, Trong-Duy Nguyen, and Masayoshi Tomizuka. The robustly-safe automated driving system for enhanced active safety. In *SAE Technical Paper*, number 2017-01-1406, 2017.

125. Changliu Liu, Chung-Wei Lin, Shinichi Shiraishi, and Masayoshi Tomizuka. Distributed conflict resolution for connected autonomous vehicles. *IEEE Transactions on Intelligent Vehicles*, 3(1):1–12, 2018.

126. Changliu Liu, Chung-Yen Lin, and Masayoshi Tomizuka. The convex feasible set algorithm for real time optimization in motion planning. *SIAM Journal on Control and Optimization*, page in review, 2016.

127. Changliu Liu, Chung-Yen Lin, and Masayoshi Tomizuka. The convex feasible set algorithm for real time optimization in motion planning. *SIAM Journal on Control and Optimization*, 56(4):2712–2733, 2018.

128. Changliu Liu, Chung-Yen Lin, Yizhou Wang, and Masayoshi Tomizuka. Convex feasible set algorithm for constrained trajectory smoothing. In *Proceedings of the American Control Conference (ACC)*, pages 4177–4182, 2017.

129. Changliu Liu, Te Tang, Hsien-Chung Lin, Yujiao Cheng, and Masayoshi Tomizuka. Serocs: Safe and efficient robot collaborative systems for next generation intelligent industrial co-robots. *arXiv preprint arXiv:1809.08215*, 2018.

130. Changliu Liu and Masayoshi Tomizuka. Control in a safe set: Addressing safety in human robot interactions. In *Proceedings of the ASME Dynamic Systems and Control Conference (DSCC)*, page V003T42A003, 2014.

131. Changliu Liu and Masayoshi Tomizuka. Modeling and controller design of cooperative robots in workspace sharing human-robot assembly teams. In *Proceedings of the IEEE/RSJ International Conference on Intelligent Robots and Systems (IROS)*, pages 1386–1391, 2014.

132. Changliu Liu and Masayoshi Tomizuka. Safe exploration: Addressing various uncertainty levels in human robot interactions. In *Proceedings of the American Control Conference (ACC)*, pages 465–470, 2015.

133. Changliu Liu and Masayoshi Tomizuka. Algorithmic safety measures for intelligent industrial co-robots. In *Proceedings of the IEEE International Conference on Robotics and Automation (ICRA)*, pages 3095–3102, 2016.

134. Changliu Liu and Masayoshi Tomizuka. Enabling safe freeway driving for automated vehicles. In *Proceedings of the American Control Conference (ACC)*, pages 3461–3467, 2016.

135. Changliu Liu and Masayoshi Tomizuka. Designing the robot behavior for safe human robot interactions. In *Trends in Control and Decision-Making for Human-Robot Collaboration Systems*, pages 241–270. Springer, 2017.

136. Changliu Liu and Masayoshi Tomizuka. Real time trajectory optimization for nonlinear robotic systems: Relaxation and convexification. *Systems & Control Letters*, 108:56–63, 2017.

137. Changliu Liu, Yizhou Wang, and Masayoshi Tomizuka. Boundary layer heuristic for search-based nonholonomic path planning in maze-like environments. In *Proceedings of the IEEE Intelligent Vehicles Symposium (IV)*, pages 831–836, 2017.

138. Changliu Liu, Wei Zhan, and Masayoshi Tomizuka. Speed profile planning in dynamic environments via temporal optimization. In *Proceedings of the IEEE Intelligent Vehicles Symposium (IV)*, pages 154–159, 2017.

139. Changliu Liu, Wenlong Zhang, and Masayoshi Tomizuka. Who to blame? Learning and control strategies with information asymmetry. In *Proceedings of the American Control Conference (ACC)*, pages 4859–4864, 2016.

140. Jiming Liu. *Autonomous Agents and Multi-Agent Systems: Explorations in Learning, Self-Organization, and Adaptive Computation*. World Scientific, 2001.

141. Xinfu Liu. *Autonomous trajectory planning by convex optimization*. PhD thesis, Iowa State University, 2013.

142. Xinfu Liu and Ping Lu. Solving nonconvex optimal control problems by convex optimization. *Journal of Guidance, Control, and Dynamics*, 37(3):750–765, 2014.

143. Yun-Hui Liu. Qualitative test and force optimization of 3-d frictional form-closure grasps using linear programming. *IEEE Transactions on Robotics and Automation*, 15(1):163–173, 1999.

144. Yun-Hui Liu. Computing n-finger form-closure grasps on polygonal objects. *The International Journal of Robotics Research*, 19(2):149–158, 2000.
145. Steve Lohr. The age of big data. *New York Times*, 11(2012), 2012.
146. Lu Lu and John T. Wen. Human-robot cooperative control for mobility impaired individuals. In *Proceedings of the American Control Conference (ACC)*, pages 447–452, 2015.
147. Ren C Luo, Han B Huang, CY Yi, and Yi W Perng. Adaptive impedance control for safe robot manipulator. In *Proceedings of the World Congress on Intelligent Control and Automation (WCICA)*, pages 1146–1151, 2011.
148. Ruikun Luo and Dmitry Berenson. A framework for unsupervised online human reaching motion recognition and early prediction. In *Proceedings of the IEEE/RSJ International Conference on Intelligent Robots and Systems (IROS)*, pages 2426–2433, 2015.
149. Vadim Macagon and Burkhard Wünsche. Efficient collision detection for skeletally animated models in interactive environments. In *Proceedings of Image and Vision Computing New Zealand (IVCNZ)*, volume 3, pages 378–383, 2003.
150. Jeffrey Mahler, Jacky Liang, Sherdil Niyaz, Michael Laskey, Richard Doan, Xinyu Liu, Juan Aparicio Ojea, and Ken Goldberg. Dex-net 2.0: Deep learning to plan robust grasps with synthetic point clouds and analytic grasp metrics. *arXiv preprint arXiv:1703.09312*, 2017.
151. Jeffrey Mahler, Florian T Pokorny, Brian Hou, Melrose Roderick, Michael Laskey, Mathieu Aubry, Kai Kohlhoff, Torsten Kröger, James Kuffner, and Ken Goldberg. Dex-net 1.0: A cloud-based network of 3d objects for robust grasp planning using a multi-armed bandit model with correlated rewards. In *Robotics and Automation (ICRA), 2016 IEEE International Conference on*, pages 1957–1964. IEEE, 2016.
152. Jim Mainprice and Dmitry Berenson. Human-robot collaborative manipulation planning using early prediction of human motion. In *Proceedings of the IEEE/RSJ International Conference on Intelligent Robots and Systems (IROS)*, pages 299–306, 2013.
153. Julie L Marble, David J Bruemmer, Douglas A Few, and Donald D Dudenhoeffer. Evaluation of supervisory vs. peer-peer interaction with human-robot teams. In *Proceedings of the Annual Hawaii International Conference on System Sciences*, pages 1–9, 2004.
154. Ian M Mitchell. Comparing forward and backward reachability as tools for safety analysis. In *International Workshop on Hybrid Systems: Computation and Control*, pages 428–443. Springer, 2007.
155. Volodymyr Mnih, Koray Kavukcuoglu, David Silver, Alex Graves, Ioannis Antonoglou, Daan Wierstra, and Martin Riedmiller. Playing atari with deep reinforcement learning. *arXiv preprint arXiv:1312.5602*, 2013.
156. Volodymyr Mnih, Koray Kavukcuoglu, David Silver, Andrei A Rusu, Joel Veness, Marc G Bellemare, Alex Graves, Martin Riedmiller, Andreas K Fidjeland, Georg Ostrovski, et al. Human-level control through deep reinforcemet learning. *Nature*, 518(7540):529–533, 2015.
157. Seyed Sajad Mousavi, Michael Schukat, and Enda Howley. Deep reinforcement learning: an overview. In *Proceedings of SAI Intelligent Systems Conference*, pages 426–440. Springer, 2016.
158. Richard M Murray, Zexiang Li, S Shankar Sastry, and S Shankara Sastry. *A mathematical Introduction to Robotic Manipulation*. CRC Press, 1994.
159. Andriy Myronenko and Xubo Song. Point set registration: Coherent point drift. *Pattern Analysis and Machine Intelligence, IEEE Transactions on*, 32(12):2262–2275, 2010.

160. Xiaoxiang Na and David J Cole. Linear quadratic game and non-cooperative predictive methods for potential application to modelling driver–AFS interactive steering control. *Vehicle System Dynamics*, 51(2):165–198, 2013.

161. Van-Duc Nguyen. Constructing force-closure grasps. *The International Journal of Robotics Research*, 7(3):3–16, 1988.

162. Stefanos Nikolaidis and Julie Shah. Human-robot cross-training: computational formulation, modeling and evaluation of a human team training strategy. In *Proceedings of the ACM/IEEE International Conference on Human-Robot Interaction*, pages 33–40, 2013.

163. Jorge Nocedal and Stephen Wright. *Numerical optimization*. Springer Science & Business Media, Springer-Verlag New York, 2006.

164. Jeanne Ellis Ormrod and Kevin M Davis. *Human learning*. Merrill London, 2004.

165. Kartik Pandit, Dipak Ghosal, H Michael Zhang, and Chen-Nee Chuah. Adaptive traffic signal control with vehicular ad hoc networks. *IEEE Transactions on Vehicular Technology*, 62(4):1459–1471, 2013.

166. Chavdar Papazov, Sami Haddadin, Sven Parusel, Kai Krieger, and Darius Burschka. Rigid 3d geometry matching for grasping of known objects in cluttered scenes. *The International Journal of Robotics Research*, 31(4):538–553, 2012.

167. Raja Parasuraman, Thomas B Sheridan, and Christopher D Wickens. A model for types and levels of human interaction with automation. *IEEE Transactions on Systems, Man, and Cybernetics-Part A: Systems and Humans*, 30(3):286–297, 2000.

168. Dae-Hyung Park, Heiko Hoffmann, Peter Pastor, and Stefan Schaal. Movement reproduction and obstacle avoidance with dynamic movement primitives and potential fields. In *Proceedings of the IEEE-RAS International Conference on Humanoid Robots*, pages 91–98, 2008.

169. Hyun-Keun Park, Hyun Seok Hong, Han Jo Kwon, and Myung Jin Chung. A nursing robot system for the elderly and the disabled. *International Journal of Human-Friendly Welfare Robotic Systems*, 2(4):11–16, 2001.

170. Vladimir Pavlovic, James M Rehg, Tat-Jen Cham, and Kevin P Murphy. A dynamic Bayesian network approach to figure tracking using learned dynamic models. In *Proceedings of the IEEE International Conference on Computer Vision (ICCV)*, volume 1, pages 94–101, 1999.

171. Leif E Peterson. K-nearest neighbor. *Scholarpedia*, 4(2):1883, 2009.

172. Lerrel Pinto and Abhinav Gupta. Supersizing self-supervision: Learning to grasp from 50k tries and 700 robot hours. In *Robotics and Automation (ICRA), 2016 IEEE International Conference on*, pages 3406–3413. IEEE, 2016.

173. Jean Ponce and Bernard Faverjon. On computing three-finger force-closure grasps of polygonal objects. *IEEE Transactions on Robotics and Automation*, 11(6):868–881, 1995.

174. Xiangjun Qian, Iñaki Navarro, Arnaud de La Fortelle, and Fabien Moutarde. Motion planning for urban autonomous driving using Bézier curves and MPC. In *Proceedings of the IEEE International Conference on Intelligent Transportation Systems (ITSC)*, pages 826–833, 2016.

175. Eric Rasmusen and Basil Blackwell. *Games and Information: An Introduction to Game Theory*. Cambridge, MA, 1994.

176. Jens Rasmussen. Outlines of a hybrid model of the process plant operator. In *Monitoring Behavior and Supervisory Control*, pages 371–383. Springer, 1976.

177. Nathan Ratliff, Matt Zucker, J Andrew Bagnell, and Siddhartha Srinivasa. CHOMP: Gradient optimization techniques for efficient motion planning. In *Proceedings of the IEEE International Conference on Robotics and Automation (ICRA)*, pages 489–494, 2009.

178. Harish C Ravichandar and Ashwin Dani. Human intention inference and motion modeling using approximate e-m with online learning. In *Proceedings of the IEEE/RSJ International Conference on Intelligent Robots and Systems (IROS)*, pages 1819–1824, 2015.

179. Harish C Ravichandar, Avnish Kumar, and Ashwin Dani. Bayesian human intention inference through multiple model filtering with gaze-based priors. In *Proceedings of the International Conference on Information Fusion (FUSION)*, pages 2296–2302, 2016.

180. John H Reif and Hongyan Wang. Social potential fields: A distributed behavioral control for autonomous robots. *Robotics and Autonomous Systems*, 27(3):171 – 194, 1999.

181. Wei Ren, Randal W Beard, et al. Consensus seeking in multiagent systems under dynamically changing interaction topologies. *IEEE Transactions on Automatic Control*, 50(5):655–661, 2005.

182. GZ Rey, M Carvalho, and D Trentesaux. Cooperation models between humans and artificial self-organizing systems: Motivations, issues and perspectives. In *Proceedings of the International Symposium on Resilient Control Systems (ISRCS)*, pages 156–161, 2013.

183. G Rigatos and S Tzafestas. Extended kalman filtering for fuzzy modelling and multi-sensor fusion. *Mathematical and Computer Modelling of Dynamical Systems*, 13(3):251–266, 2007.

184. Philip E Ross. California to issue driving licences to robots. *IEEE Spectrum*, 2014.

185. Dorsa Sadigh, Shankar Sastry, Sanjit A Seshia, and Anca D Dragan. Planning for autonomous cars that leverages effects on human actions. In *Proceedings of the Robotics: Science and Systems Conference (RSS)*, 2016.

186. Andrey V Savkin, Alexey S Matveev, Michael Hoy, and Chao Wang. *Safe Robot Navigation Among Moving and Steady Obstacles*. Butterworth-Heinemann, 2015.

187. Stefan Schaal, Auke Ijspeert, and Aude Billard. Computational approaches to motor learning by imitation. *Philosophical Transactions of the Royal Society B: Biological Sciences*, 358(1431):537–547, 2003.

188. John Schulman, Yan Duan, Jonathan Ho, Alex Lee, Ibrahim Awwal, Henry Bradlow, Jia Pan, Sachin Patil, Ken Goldberg, and Pieter Abbeel. Motion planning with sequential convex optimization and convex collision checking. *The International Journal of Robotics Research*, 33(9):1251–1270, 2014.

189. John Schulman, Jonathan Ho, Alex X Lee, Ibrahim Awwal, Henry Bradlow, and Pieter Abbeel. Finding locally optimal, collision-free trajectories with sequential convex optimization. In *Proceedings of the Robotics: Science and Systems Conference (RSS)*, volume 9, pages 1–10, 2013.

190. John Schulman, Jonathan Ho, Cameron Lee, and Pieter Abbeel. Learning from demonstrations through the use of non-rigid registration. In *Proceedings of the 16th International Symposium on Robotics Research (ISRR)*, 2013.

191. Mac Schwager, Carrick Detweiler, Iuliu Vasilescu, Dean M Anderson, and Daniela Rus. Data-driven identification of group dynamics for motion prediction and control. *Journal of Field Robotics*, 25(6-7):305–324, 2008.

192. Loren Arthur Schwarz, Artashes Mkhitaryan, Diana Mateus, and Nassir Navab. Human skeleton tracking from depth data using geodesic distances and optical flow. *Image and Vision Computing*, 30(3):217–226, 2012.

193. Elham Semsar-Kazerooni and Khashayar Khorasani. Multi-agent team cooperation: A game theory approach. *Automatica*, 45(10):2205 – 2213, 2009.

194. David Silver, Aja Huang, Chris J Maddison, Arthur Guez, Laurent Sifre, George Van Den Driessche, Julian Schrittwieser, Ioannis Antonoglou, Veda Panneershelvam, Marc Lanctot, et al. Mastering the game of go with deep neural networks and tree search. *Nature*, 529(7587):484–489, 2016.

195. Emrah A Sisbot, Luis Felipe Marin-Urias, Xavier Broquere, Daniel Sidobre, and Rachid Alami. Synthesizing robot motions adapted to human presence. *International Journal of Social Robotics*, 2(3):329–343, 2010.

196. S Skiena. Dijkstra's algorithm. *Implementing Discrete Mathematics: Combinatorics and Graph Theory with Mathematica, Reading, MA: Addison-Wesley*, pages 225–227, 1990.

197. Jean-Jacques E Slotine, Weiping Li, et al. *Applied nonlinear control*, volume 199. Prentice-Hall Englewood Cliffs, NJ, 1991.

198. Peter Spellucci. A new technique for inconsistent QP problems in the SQP method. *Mathematical Methods of Operations Research*, 47(3):355–400, 1998.

199. Roman G Strongin and Yaroslav D Sergeyev. *Global optimization with non-convex constraints: Sequential and parallel algorithms*, volume 45. Springer Science & Business Media, 2013.

200. Liting Sun, Cheng Peng, Wei Zhan, and Masayoshi Tomizuka. A fast integrated planning and control framework for autonomous driving. In *arXiv:1707.02515*, 2017.

201. Wen Sun, Arun Venkatraman, Geoffrey J Gordon, Byron Boots, and J Andrew Bagnell. Deeply aggrevated: Differentiable imitation learning for sequential prediction. In *Proceedings of the 34th International Conference on Machine Learning-Volume 70*, pages 3309–3318. JMLR. org, 2017.

202. Supplementary website for Te Tang's dissertation. `http://me.berkeley.edu/~tetang/Dissertation`.

203. Susumu Tachi and Kiyoshi Komoriya. Guide dog robot. *Autonomous Mobile Robots: Control, Planning, and Architecture*, pages 360–367, 1984.

204. Tadele S Tadele, Theo JA de Vries, and Stefano Stramigioli. The safety of domestic robots: a survey of various safety-related publications. *IEEE Robotics and Automation Magazine*, pages 134–142, 2014.

205. JTC Tan, F Duan, Y Zhang, K Watanabe, R Kato, and T Arai. Human-robot collaboration in cellular manufacturing: Design and development. In *Proceedings of the IEEE/RSJ International Conference on Intelligent Robots and Systems (IROS)*, pages 29–34, 2009.

206. Te Tang, Hsien-Chung Lin, and Masayoshi Tomizuka. A learning-based framework for robot peg-hole-insertion. In *Proceedings of the ASME Dynamic Systems and Control Conference (DSCC)*, pages V002T27A002–V002T27A002, 2015.

207. Te Tang, Hsien-Chung Lin, Yu Zhao, Wenjie Chen, and Masayoshi Tomizuka. Autonomous alignment of peg and hole by force/torque measurement for robotic assembly. In *Proceedings of the IEEE International Conference on Automation Science and Engineering (CASE)*, pages 162–167, 2016.

208. Te Tang, Hsien-Chung Lin, Yu Zhao, Yongxiang Fan, Wenjie Chen, and Masayoshi Tomizuka. Teach industrial robots peg-hole-insertion by human demonstration. In *Proceedings of the IEEE International Conference on Advanced Intelligent Mechatronics (AIM)*, pages 488–494, 2016.

209. Te Tang, Changliu Liu, Wenjie Chen, and Masayoshi Tomizuka. Robotic manipulation of deformable objects by tangent space mapping and non-rigid registration. In *Proceedings of the IEEE/RSJ International Conference on Intelligent Robots and Systems (IROS)*, pages 2689–2696, 2016.

210. Mohit Tawarmalani and Nikolaos V Sahinidis. *Convexification and Global Optimization in Continuous and Mixed-Integer Nonlinear Programming: Theory, Algorithms, Software, and Applications*, volume 65. Springer Science & Business Media, 2002.

211. Andrea Lockerd Thomaz, Cynthia Breazeal, et al. Reinforcement learning with human teachers: Evidence of feedback and guidance with implications for learning performance. In *Proceedings of the National Conference on Artificial Intelligence (AAAI)*, volume 6, pages 1000–1005, 2006.

212. Sebastian Thrun and Tom M Mitchell. Lifelong robot learning. In *The Biology and Technology of Intelligent Autonomous Agents*, pages 165–196. Springer, 1995.

213. Kaoru Tone. Revisions of constraint approximations in the successive QP method for nonlinear programming problems. *Mathematical Programming*, 26(2):144–152, 1983.

214. Giovanni Tonietti, Riccardo Schiavi, and Antonio Bicchi. Design and control of a variable stiffness actuator for safe and fast physical human/robot interaction. In *Proceedings of the IEEE International Conference on Robotics and Automation (ICRA)*, pages 526–531, 2005.

215. Peter Trautman and Andreas Krause. Unfreezing the robot: Navigation in dense, interacting crowds. In *Proceedings of the IEEE/RSJ International Conference on Intelligent Robots and Systems (IROS)*, pages 797–803, 2010.

216. Chi-Shen Tsai, Jwu-Sheng Hu, and Masayoshi Tomizuka. Ensuring safety in human-robot coexistence environment. In *Proceedings of the IEEE/RSJ International Conference on Intelligent Robots and Systems (IROS)*, pages 4191–4196, 2014.

217. Annemarie Turnwald, Wiktor Olszowy, Dirk Wollherr, and Martin Buss. Interactive navigation of humans from a game theoretic perspective. In *Proceedings of the IEEE/RSJ International Conference on Intelligent Robots and Systems (IROS)*, pages 703–708, 2014.

218. Gündüz Ulusoy, Funda Sivrikaya-Şerifoğlu, and Ümit Bilge. A genetic algorithm approach to the simultaneous scheduling of machines and automated guided vehicles. *Computers and Operations Research*, 24(4):335–351, 1997.

219. Chris Urmson, Joshua Anhalt, Drew Bagnell, Christopher Baker, Robert Bittner, MN Clark, John Dolan, Dave Duggins, Tugrul Galatali, Chris Geyer, et al. Autonomous driving in urban environments: Boss and the urban challenge. *Journal of Field Robotics*, 25(8):425–466, 2008.

220. Jur van den Berg. Extended LQR: locally-optimal feedback control for systems with non-linear dynamics and non-quadratic cost. In *Robotics Research*, pages 39–56. Springer, 2016.

221. Jur Van Den Berg, Pieter Abbeel, and Ken Goldberg. LQG-MP: Optimized path planning for robots with motion uncertainty and imperfect state information. *The International Journal of Robotics Research*, 30(7):895–913, 2011.

222. Timothy Van Zandt. Interim bayesian nash equilibrium on universal type spaces for supermodular games. *Journal of Economic Theory*, 145(1):249–263, 2010.

223. Jacob Varley, Jonathan Weisz, Jared Weiss, and Peter Allen. Generating multi-fingered robotic grasps via deep learning. In *Intelligent Robots and Systems (IROS), 2015 IEEE/RSJ International Conference on*, pages 4415–4420. IEEE, 2015.

224. Andreas Vlachos. An investigation of imitation learning algorithms for structured prediction. In *European Workshop on Reinforcement Learning*, pages 143–154, 2013.

225. Michael Wooldridge. *An Introduction to Multiagent Systems*. John Wiley & Sons, 2009.

226. Ping Wu, Yang Cao, Yuqing He, and Decai Li. Vision-based robot path planning with deep learning. In *International Conference on Computer Vision Systems*, pages 101–111. Springer, 2017.

227. Wenda Xu, Junqing Wei, J.M. Dolan, Huijing Zhao, and Hongbin Zha. A real-time motion planner with trajectory optimization for autonomous vehicles. In *Proceedings of the IEEE International Conference on Robotics and Automation (ICRA)*, pages 2061–2067, 2012.

228. Chenguang Yang, Gowrishankar Ganesh, Sami Haddadin, Sven Parusel, Alin Albu-Schaeffer, and Etienne Burdet. Human-like adaptation of force and impedance in stable and unstable interactions. *IEEE Transactions on Robotics*, 27(5):918–930, 2011.

229. Shichao Yang, Sandeep Konam, Chen Ma, Stephanie Rosenthal, Manuela Veloso, and Sebastian Scherer. Obstacle avoidance through deep networks based intermediate perception. *arXiv preprint arXiv:1704.08759*, 2017.

230. James E Young, Richard Hawkins, Ehud Sharlin, and Takeo Igarashi. Toward acceptable domestic robots: Applying insights from social psychology. *International Journal of Social Robotics*, 1(1):95–108, 2009.

231. Wenlong Zhang, Xu Chen, Joonbum Bae, and Masayoshi Tomizuka. Real-time kinematic modeling and prediction of human joint motion in a networked rehabilitation system. In *American Control Conference (ACC)*, pages 5800–5805, 2015.

232. Yu Zhang, Huiyan Chen, Steven L Waslander, Tian Yang, Sheng Zhang, Guangming Xiong, and Kai Liu. Speed planning for autonomous driving via convex optimization. In *2018 21st International Conference on Intelligent Transportation Systems (ITSC)*, pages 1089–1094. IEEE, 2018.

233. Zhenyue Zhang and Jing Wang. Mlle: Modified locally linear embedding using multiple weights. In *Advances in Neural Information Processing Systems*, pages 1593–1600, 2007.

234. Zhijie Zhu, Edward Schmerling, and Marco Pavone. A convex optimization approach to smooth trajectories for motion planning with car-like robots. In *Proceedings of the IEEE Conference on Decision and Control (CDC)*, pages 835–842, 2015.

235. Michael Zinn, Bernard Roth, Oussama Khatib, and J Kenneth Salisbury. A new actuation approach for human friendly robot design. *The International Journal of Robotics Research*, 23(4-5):379–398, 2004.

Index

a posteriori, 86
a priori, 86

admittance, 92
    admittance controller, 92
agent, 27
analogy learning, 97
    analogy, 97
    structure preserved registration, 103
asymptotically stable, 94

behavior system, 7
    knowledge, 11
        cost, 12
        model, 13
        learning, 16
        logic, 14
belief space, 54

Cartesian space, 63
chance constraint, 55
Chicken game, 134
co-robot, 3
configuration space, 62
conflict graph, 140
conflict resolution, 121
convex feasible set algorithm, 19
convexification, 18

deadlock, 120, 131
design objectives, 17
    cooperation, 20
    dexterity, 20
    efficiency, 18
    imitation, 19
    safety, 17
dynamics
    steady-state response, 120
    transient response, 120

evaluation platform, 22

Gaussian mixture regression, 90

grasp planning
    dissimilarity, 185
    grasping pose, 183
    grasping quality, 182
hidden Markov model, 84
human-robot interaction, 3

information structure, 30
    complete information, 29, 122
    incomplete information, 29, 122
    information asymmetry, 30, 119, 122
invariance principle
    LaSalle's, 95

mean squared estimation error, 54
model predictive control, 33
modes of interaction, 4
    hierarchical relationship, 6, 30
    parallel relationship, 5, 30
    space-sharing, 29
    time-sharing, 29
modes of interactions
    space-sharing, 149
motion planning
    optimization-based, 18
    spatiotemporal planning, 61
motion re-planning, 192
    initial reference, 193
    path re-replanning, 193
    trajectory re-planning, 195
multi-agent system
    conflict of interests, 45
    conflict resolution, 45, 121
    equilibrium, 120
    sequential game, 30
    simultaneous game, 30, 121
    time-varying topology, 30

potential field algorithm, 18

recursive least square, 87, 128

remote lead through teaching, 175
    human demonstration device, 175

safe exploration algorithm, 46, 54
safe set algorithm, 46, 49
safety
    interactive safety, 45
    intrinsic safety, 45
    safety index, 47
    safety principle, 47
semidefinite programming, 18
sequential quadratic programming, 18
skew symmetry, 95
slack convex feasible set algorithm, 19,
    75
sliding mode algorithm, 18
social norm, 45
speed profile, 63, 72

time invariance, 47
trapped equilibrium, 124

uncertainty
    accumulation of uncertainty, 33
    structural uncertainty, 31
    uncertainty cone, 33
    uncertainty reduction, 45

Color Section

Chapter 1: Fig. 1.1, p. 5

Collaboration   Competition

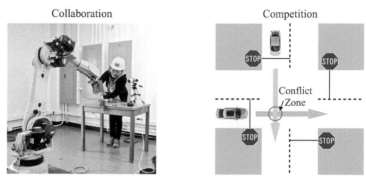

(a) Parallel relationships in human-robot interactions.

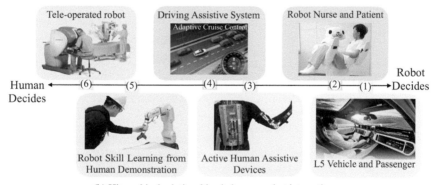

Tele-operated robot   Driving Assistive System   Robot Nurse and Patient

Human ←——(6)————(5)————————(4)——(3)———————(2)—(1)→ Decides
Decides   Robot

Robot Skill Learning from   Active Human Assistive   L5 Vehicle and Passenger
Human Demonstration   Devices

(b) Hierarchical relationships in human-robot interactions.

Chapter 2: Fig. 2.7, p. 37

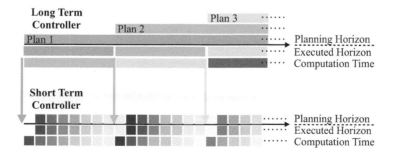

Chapter 8: Fig. 8.1, p. 150

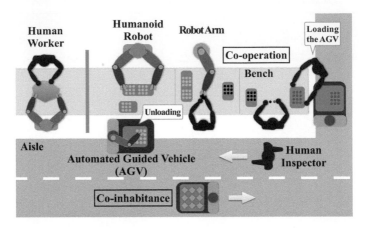

Chapter 9: Fig. 9.19, p. 197

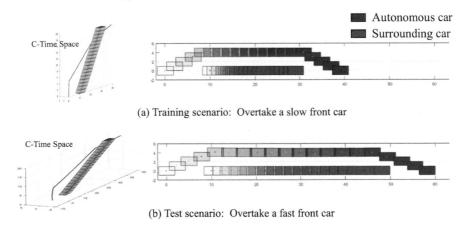

(a) Training scenario: Overtake a slow front car

(b) Test scenario: Overtake a fast front car

Chapter 9: Fig. 9.20, p. 197

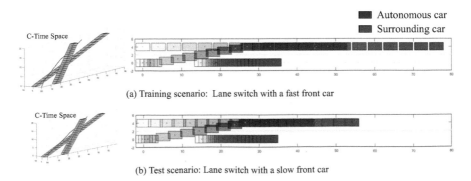

(a) Training scenario: Lane switch with a fast front car

(b) Test scenario: Lane switch with a slow front car

Chapter 9: Fig. 9.21, p. 198

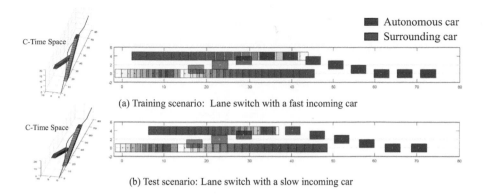

(a) Training scenario: Lane switch with a fast incoming car

(b) Test scenario: Lane switch with a slow incoming car

Chapter 10: Fig. 10.3, p. 206

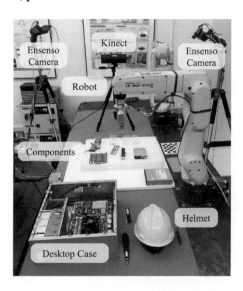

Chapter 10: Fig. 10.5, p. 208

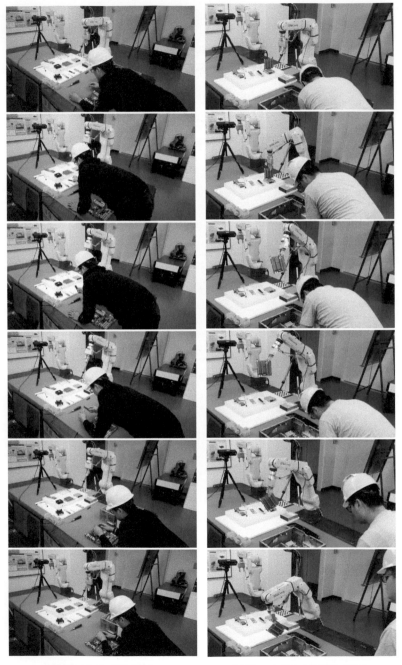

(a) Robot in the Idle Mode.   (b) Human-Robot Collaboration.

Chapter 10: Fig. 10.6, p. 209

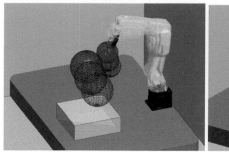
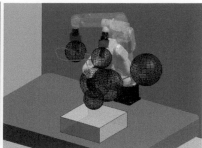

(a) Robot in the Idle Mode.   (b) Human-Robot Collaboration.